British Literature and Technology, 1600–1830

Aperçus: Histories Texts Cultures

Series editor: Kat Lecky, Loyola University Chicago

Aperçus is a series of books exploring the connections among historiography, culture, and textual representation in various disciplines. Revisionist in intention, *Aperçus* seeks monographs as well as guest-edited multiauthored volumes that stage critical interventions to open up new possibilities for interrogating how systems of knowledge production operate at the intersections of individual and collective thought.

We are particularly interested in medieval, Renaissance, early modern, and Restoration texts and contexts. Areas of focus include premodern conceptions and theorizations of race, ethnicity, gender, and sexuality in art, literature, historical artefacts, medical and scientific works, political tracts, and religious texts; negotiations between local, national, and imperial intellectual spheres; the cultures, literatures, and politics of the excluded and marginalized; print history and the history of the book; the medical humanities; and the cross-pollination of humanistic and scientific modes of inquiry.

Recent titles in the series:

British Literature and Technology, 1600–1830
Kristin M. Girten and Aaron R. Hanlon, eds.

Deep Mapping the Literary Lake District: A Geographical Text Analysis
Joanna E. Taylor and Ian N. Gregory

Oriental Networks: Culture, Commerce, and Communication in the Long Eighteenth Century
Greg Clingham and Bärbel Czennia, eds.

Writing Lives in the Eighteenth Century
Tanya Caldwell, ed.

For more information about the series, please visit www.bucknelluniversitypress.org.

British Literature and Technology, 1600–1830

~

Edited by Kristin M. Girten
and Aaron R. Hanlon

Bucknell University Press
Lewisburg, Pennsylvania

978-1-68448-396-9 (cloth)
978-1-68448-395-2 (paper)
978-1-68448-397-6 (epub)

Cataloging-in-publication data is available from the Library of Congress.
LCCN 2022010612

A British Cataloging-in-Publication record for this book is available from the British Library.

∞ The paper used in this publication meets the requirements of the American National Standard for Information Sciences—Permanence of Paper for Printed Library Materials, ANSI Z39.48-1992.

www.bucknelluniversitypress.org

Distributed worldwide by Rutgers University Press

Printed and bound by CPI Group (UK) Ltd, Croydon, CR0 4YY

Contents

British Literature and Technology, 1600–1830

Introduction

Kristin M. Girten and Aaron R. Hanlon

What can a poem accomplish? In the case of Anna Letitia Barbauld's "The Mouse's Petition" (1773), it can save a mouse from a deadly experimental technology. According to the memoirist William Turner, who knew both Barbauld and Joseph Priestley, Barbauld became aware that Priestley had caught a mouse to put to scientific purpose. In a series of experiments that would lead to his discovery of oxygen, Priestley was studying the effects of pumping different gases into a sealed chamber to commingle with air and would conduct such experiments with a live mouse in the chamber. It was too late in the evening for Priestley to undertake the experiment, so he asked a servant to put the mouse aside for the next day, an exchange Barbauld witnessed. That night she composed the poem in the plaintive voice of the mouse and left the paper bearing the poem "twisted among the wires of the cage," as Turner reports, for Priestley to find. The next day Priestley found the poem and freed the mouse.[1]

In this example, we have a literary act (Barbauld's poem) intervening in a scientific and technological act (Priestley's experiment), although these labels suggest a greater degree of distinction between literature, science, and technology than is realistic. In their landmark study *Leviathan and the Air-Pump: Hobbes, Boyle, and the Experimental Life*, Steven Shapin and Simon Schaffer use the phrase "literary technology" to describe one of the three technologies that seventeenth-century experimental natural philosophers used to produce and establish matters of fact. Taking as an illustrative example Robert Boyle's air-pump—a machine that produced a vacuum by drawing air out of a sealed chamber, not unlike what Priestley would use a hundred years later in his oxygen experiments—Shapin and Schaffer identify "literary technology" as the means by which "the phenomena produced by the pump were made known to those who were not direct witnesses." Because the establishment of fact in the seventeenth- and eighteenth-century Royal Society of London for Improving Natural Knowledge paradigm is "both an epistemological and a social" matter, such that experimental results needed to be demonstrated and communicated, the means by which experimental natural philosophers achieved this could be understood as *technologies*. That is, what Shapin and Schaffer call "material," "literary," and "social" technologies are "*knowledge-producing tools*" that, like Barbauld's poem, *make something happen*.[2] They do not constitute scientific practice in and of themselves; they enable it. For the purposes of this volume, we understand technology as a tool—whether physical or conceptual—but not necessarily a tool over which we humans have complete control. That is, we proceed from Andrew Feenberg's view that the idea of technology as "an alien force intruding on our social life from a coldly rational beyond" is illusory.[3] In the Baconian sense, technologies always mediate between us and our surroundings, but mediation runs in both directions: we use tools to shape ourselves and our surroundings, and the tools we use come to shape us as well. As Neil Rhodes and Jonathan Sawday demonstrate, the impact of the printing press on the sixteenth century ("the first century of print") is analogous to the situation "facing publishers and readers in our own period, the first century of computers," and in both scenarios technology is not a passive tool but an active force in shaping social, economic, and political life.[4]

"Literary technology," then, is "literary" in the sense that it "dramatized" through "literary exposition" the "forms of social organization" societies of experimental natural philosophers took as well as the results of their experiments.[5] Shapin and Schaffer call the literary technology that Boyle employed in advertising his air-pump "virtual witnessing," because it meant coming up with ways to describe the material technologies of experiments (the air-pump itself), the accompanying procedures and social arrangements in experimentation, and the observed results in ways that would earn trust in the integrity of the process among those who could not be present to see the experiment with their own eyes.[6] As more recent scholarship on this topic has shown—particularly Tita Chico's *The Experimental Imagination: Literary Knowledge and Science in the British Enlightenment*—both the linguistic features common in literary texts and literary texts themselves—poems, novels, and plays—were integral to the description and conceptual formation of scientific work in the period. That is, literary technologies included not only the narrative adventures and metaphorical portrayals common in scientific atlases such as Robert Hooke's *Micrographia* (1665) but also fiction such as Margaret Cavendish's *The Blazing World* (1666) and Jonathan Swift's *Gulliver's Travels* (1726), poems such as Abraham Cowley's "Ode to the Royal Society" (1667) and Alexander Pope's *Essay on Man* (1733–1734), and plays such as Thomas Shadwell's *The Virtuoso* (1676). Not only did such texts represent and critique experimental science; they also supplied the literary technologies that made science possible.

What such use of literary technologies suggests—as Shapin and Schaffer's concept of "literary technology" already captures—is that literature itself can function as technology.[7] This insight, from Shapin and Schaffer to more recent studies by Joanna Picciotto, Courtney Weiss Smith, Joseph Drury, Al Coppola, Tita Chico, and others, has allowed literature scholars and historians of science together to achieve a profound understanding of how early modern and Enlightenment science worked.[8] Accordingly, studies in which literature and technology appear together as objects of inquiry have tended to take the understanding of literature and science as animating goals, toward which understanding literature *as* technology has been fruitful. This volume makes contributions to that mission, but it also does something different. In this volume, we understand not only literature *as* technology but also literature *and* technology, or the ways literary texts engage with other technologies—the textile loom, the steam engine, the telegraph—not only in representing them but also in shaping their cultural meanings and functions. In *Questioning Technology*, Andrew Feenberg emphasizes technology's "ambivalence": its "availability . . . for alternative developments with different social consequences."[9] Our volume shows that Enlightenment authors were readily aware of such ambivalence. In fact, in many cases, they influenced technological consequences (and in some instances technological developments) by influencing technologies' reception. As Feenberg explains, "broad society-wide factors . . . shape technology behind the backs of the actors."[10] Writing at the dawn of the Industrial Age, the authors whom our contributors discuss actively participate in such "shaping." By witnessing their interventions in emerging technological developments, our volume challenges the cultural phenomenon whereby "technology rigidifies into necessity" and thereby refutes technology's autonomy. Like history, technology is made, not born. It is, in Feenberg's words, "the medium of daily life in modern societies," continually developing in concert with social, civic, and economic life, from quotidian agricultural practices to transatlantic communication.[11] Technologies are far more than the sum of their gears and screws. Embedded within them is a diverse history of "tinkering" performed not only by the scientists who apply their technical knowledge to physical matter but also by writers working across genres and employing underacknowledged forms of *technical* expertise: choosing the right metaphor, plotting speculative fictional narratives, or making close observations of a given technology's operation and impact in the social world.

"If science and technology are intimately related concepts," asks Jennifer L. Lieberman, "why pull technology out as a subject that might be studied with literature but separately from science?"[12] Lieberman's question raises two considerations, both essential to this volume. One is about the history of knowledge organization, in particular the structure of science studies as a field or cluster of fields and the influence of this structure on the proportion of research that focuses on science versus technology. Among historians, the study of science and of technology has been divided into distinct professional societies with their own journals.[13] D.S.L. Cardwell attributes this separation to "the powerful influence of certain philosophers of science, in particular those brought up in that mainland European tradition that emphasized the distinction between the 'pure' science of the universities and the other sort, as practised and taught in the technical high schools."[14] In contrast to the history of science centered on "pure" or basic science, the history of technology, as Cardwell notes, "was popularized [in the nineteenth century] by Samuel Smiles who celebrated the triumph of the hero-engineer," the builder of "roads, canals, railways, harbours," and the like.[15] Further, as Paul

Forman argues, the primacy that modernity assigned to science (of the sort Cardwell calls "pure" science) over technology led historians of technology, when the discipline became institutionalized in the mid-twentieth century, to "keep science out—and, where it could not be kept out, to put *it* down."[16] This is not necessarily the attitude that governs the relationship between the history of science and the history of technology today, though it helps explain why these intimately related histories sometimes sit uneasily with each other.

The second consideration that Lieberman's question raises is both organizational and epistemological, the question of what has been the relationship between science and technology and to what extent it is possible or valuable to distinguish between the two. *Techne* has always implied a distinction between making or doing and understanding, or a distinction between *knowledge-how* and *knowledge-that*. Yet Enlightenment natural philosophers—particularly those of the Royal Society—frequently collapsed this distinction by emphasizing what Robert Hooke called "useful knowledge" in his preface to *Micrographia* (1665).[17] In this sense, it would seem impossible to extricate seventeenth- and eighteenth-century science and technology from each other. For this reason, our focus on literature and technology in this volume should be understood as a heuristic shift in emphasis, as opposed to an effort to instantiate a false division. What Lieberman suggests might be a valuable "separation" of science and technology is for us not a matter of separation but heuristic emphasis, given the ways the study of literature and technology has been subsumed under the study of literature and science.

Even within academic journals that combine the study of science, technology, and literature, the imbalance of "science" versus "technology" coverage is stark. As Lieberman notes, "the subject search terms *literature* and *technology* call up sixteen articles in *Configurations*, compared to the fifty-nine articles retrieved by the subject search terms *literature* and *science*.[18] If we turn more expressly to literary studies per se, a comparable search in the *MLA International Bibliography* retrieves 4,340 articles on the subjects "literature and technology" and 11,225 articles on "literature and science." Such metrics have obvious limitations—for example, articles that discuss literature and science may also discuss technology implicitly—yet they offer a rough impression of the relative visibility of "science" and "technology" as concepts within literary studies and science studies.

That eighteenth-century philosophers themselves readily acknowledged, and indeed regularly celebrated, the link between science and technology (which at the time were designated "natural philosophy" and "practical arts") makes the current tendency of literature scholars to emphasize the former while downplaying, if not completely excluding, the latter particularly detrimental. Formal societies designed to foster and exhibit innovative practical applications of scientific discoveries proliferated throughout the eighteenth century.[19] The "progress of knowledge" that Bacon and his disciples pursued was far from purely philosophical in nature. A cornerstone of Bacon's inductive empirical method was its close identification with embodied experience. Not only did experience *inform* this widely adopted method, however; it was also perceived to reap the harvest of the new method. As Bacon explains, "We are laying the foundations not of a sect or of a dogma, but of human progress and empowerment."[20] For Bacon and the Royal Society that would take up his mantle, the "Great Instauration" entailed not simply a "progress of knowledge" but a progress of civilization, a progress of humanity. In coming to know nature, we would come to know how to manipulate it for human benefit and thus ensure "the victory of art over nature."[21] In other words, the practical application of knowledge was at least as important as, if not more than, a conception of pure knowledge without use. Knowledge was not pursued simply for knowledge's sake. It was pursued out of a conviction of its practical—and thus technological—value.

Indeed, technological innovation was a hallmark of the progress envisioned and pursued by Enlightenment-era men of science. According to David Spadafora, "for the most part, the adherents of the moderns were triumphant and their belief in progress in the arts and sciences prevailed."[22] Advances in technology were instrumental in providing evidence of, and thus inspiring faith in, such progress. Bacon's editor, Lisa Jardine, suggests that those who would offer financial support to Bacon and his followers viewed such advances as of greater importance than the acquisition of knowledge itself: "The kinds of investors [Bacon] seeks for his *Great Instauration* need to know they can expect to make a rapid profit. . . . [T]hey will need to see immediate pay-off in the form of enhanced procedures in traditional trade and manufacture."[23] The main goal of the London Society of Arts was, as its founder, William Shipley, explains, "to embolden enterprise, to enlarge Science, to refine Art, to improve our Manufactures, and extend our Commerce; in a word, to render Great Britain the school of instruction, as it is already the centre of traffic to the greatest part of the known world."[24] Similar ambitions informed the many other "improving" societies founded across England, Scotland, and Ireland during the historical Enlightenment.[25] To consult the writings and proceedings of such societies is to develop a sense of the profound optimism that many men of science came to feel as a result of their collaborative endeavors. It is also to appreciate the emphasis these men placed on practical applications of science that produced technological advancements, particularly when such advancements benefited industry and, thus, "commerce." Enthusiasm for science and technology went hand in hand with national pride. According to Shipley and many other innovators like him, new technological developments could even contribute to the imagining, building, and exporting of the nation. These developments played an important role in the story of progress that British "Moderns" like Shipley sought to tell and pursue.

However, not everyone shared Shipley's optimistic view of technological innovation. Satirists regularly targeted the new science, and as they did so, modern technologies regularly came under fire. Jonathan Swift's *Gulliver's Travels* offers a particularly memorable example. Both its Brobdingnag and Laputa voyages prominently target the faith and reverence that modern natural philosophers invested in the technologies that their scientific practices employed and often helped to popularize.[26] For example, in a conversation about politics and civil affairs, the King of Brobdingnag is "struck with Horror at the Description" Gulliver has just given of England's use of gunpowder-fueled firearms to settle international disputes.[27] While Gulliver conveys great patriotic pride in England's use of this centuries' old "Invention," the King is appalled at how Gulliver can "appear wholly unmoved at all the Scenes of Blood and Desolation, which [he] had painted as the common Effects of those destructive Machines."[28] Swift's satire encourages readers to side with the King when he remarks that "some evil Genius Enemy to Mankind, must have been the first Contriver" of firearm technologies.[29] Here as elsewhere in the work, modern England appears uncivilized and barbaric, in spite of Gulliver's best attempts to garner appreciation for its modern innovations, technological and otherwise.

The early Enlightenment writer and philosopher Margaret Cavendish also challenges enthusiasm for modern technological advances—specifically, observational technologies such as the microscope and the telescope. She asserts, "Wherefore those that invented microscopes, and such like dioptrical glasses, at first, did, in my opinion, the world more injury than benefit."[30] According to Cavendish, in a chapter titled "Of Micrography, and of Magnifying and Multiplying Glasses," though experimentalists regularly employed modern observational technologies to expand their capacity for perception, such technologies were distorting rather than revealing: "Magnifying, multiplying, and the like optic glasses, may, and do oftentimes present falsely the picture of an exterior object; I say, the picture, because it is not the real body of the object which the glass presents but the glass only figures or patterns out the picture presented in and by the glass, and there mistakes may easily be committed in taking copies from copies."[31] Cavendish maintains that, far from ensuring a detached and thus reliable access to specimens as experimentalists perceived them to do, microscopes manipulate and therefore present a highly misleading and unreliable mediated depiction of them. She goes on to say that the "art" of using "dioptrical glasses . . . has intoxicated so many men's brains, and wholly employed their thoughts and bodily actions about phenomena, or the exterior figures of objects, as all better arts and studies are laid aside."[32] Both Swift and Cavendish perceive enthusiasts for modern technology as gullible, deluded, and even childish:

> As boys that play with watery bubbles or fling dust into each other's eyes, or make a hobbyhorse of snow, are worthy of reproof rather than praise, for wasting their time with useless sports, so those that addict themselves to unprofitable arts, spend more time than they reap benefit thereby. . . . Nay, could they benefit men either in husbandry, architecture, or the like necessary and profitable employments; yet before the vulgar sort would learn to understand them, the world would want bread to eat, and houses to dwell in, as also clothes to keep them from the inconveniences of the inconstant weather.[33]

For critics of modern technology such as Cavendish and Swift, zeal for innovative technologies ("unprofitable arts"), which became widespread during the historical Enlightenment, had the effect of distracting natural philosophers from finding solutions to basic problems. They portrayed technological zealots much like Swift portrays the Philosopher at the end of his "The Mechanical Operation of the Spirit": "while his Thoughts and Eyes were fixed upon the *Constellations*, [he] found himself seduced by his *lower Parts* into a *Ditch*."[34] The promise of modern technology was great but, according to some detractors, it was exaggerated and could even prove treacherous.

Focusing on convergences between literature and technology during the historical Enlightenment (and immediately before and after it), the essays in our collection register the important role writers in different genres played in mediating Enlightenment technologies and progress. Attending to lesser-known works, or pursuing new approaches to well-known ones, the following essays help us expand our appreciation of how tenuous the faith in a progressive Enlightenment narrative could be. Indeed, many of the essays in the collection portray how authors developed biting critiques of this narrative and the technological innovations that informed it. Analyzed through the framework of Enlightenment technology and progress, and understood in this volume as technologies in and of themselves, familiar works such as Daniel Defoe's *Robinson Crusoe* and Jonathan Swift's *Gulliver's Travels* reveal themselves in new ways.

Although we have structured the volume in chronological order, various themes emerge across the volume that present other ways of categorizing the chapters and of understanding their relationship to one another. For instance, chapters 1 and 2, by Laura Francis and Erik L. Johnson, respectively, emphasize literature's function *as a technology*. Chapters 3 through 7, by Thomas A. Oldham, Zachary M. Mann, Kevin MacDonnell, Emily M. West, and Deven M. Parker, highlight how literary works engaged specific contemporaneous technologies, such as the automated loom, the steam engine, the telegraph, and others. The volume concludes with an essay by Jamison Kantor in which technology functions somewhat differently than it does in the other chapters—as a metaphor that helps us understand the concept of "political machinery."

Our collection begins with Laura Francis's case study of John Webster's *The Duchess of Malfi*, which sets an important foundation for the collection as a whole as it demonstrates that British literature of the long Enlightenment era did not simply regularly depict new technologies but could also serve as a technology in and of itself. According to Francis, with *The Duchess of Malfi*, Webster employs theater as an instrument for submitting innovations of the new science to public scrutiny and evaluation, ultimately inviting his audience to develop a healthy skepticism not only about the reliability of empirical instruments (such as the compass) but also about sense perception in general. Francis's analysis demonstrates that literary authors joined natural philosophers in "shaping the technological and epistemological shifts of their time."

In chapter 2, Erik L. Johnson develops a novel approach to *Robinson Crusoe* as he pursues connections between Mary Hearne's and Daniel Defoe's depictions of time. According to Johnson, both authors pose a similar challenge to "the modern, technologically mediated experience of time," which coincided with, and helped to reinforce, John Locke's "durational model of experience" and the consistency of time it implied. Johnson argues that Hearne and Defoe appeal to an older Augustinian notion of time that is based on an analogy between the present and the eternal and that, in contrast to Locke, perceives time as relative and variable. Designed to register time's consistency, horological technologies had become increasingly valued and available over the first couple of decades of the eighteenth century. Johnson shows that Hearne and Defoe qualify the value of these technologies, suggesting that they misrepresent time by "homogenizing" it. Thus, they implicitly challenge the progress that horological innovations were presumed to enable, introducing the work of Hearne and Defoe as narrative technologies.

With chapter 3 by Thomas A. Oldham, our volume begins to consider how authors directly addressed and sometimes shaped specific technologies and their reception. Oldham turns our attention to the 1717 Scriblerian satire *Three Hours after Marriage*—a collaboration between John Gay, Alexander Pope, and John Arbuthnot. He argues that the play directs its critique at new gynecological technologies and, in doing so, undercuts the control over female bodies that such technologies were widely presumed to enable. According to Oldham, it emphasizes "the performativity of all women" and, simultaneously, the incapacity of controlling technologies to penetrate such performance. However, he also shows how the play documents "the debasement of female agency" that accompanied the employment of such technologies: though the technologies, and the knowledges that inform and/or result from them, may be laughable, their effects are nevertheless real and profoundly felt by the women who are their targets. Ultimately, women are shown within the play to be reduced to commodities, "a mere exchangeable set of goods."

In chapter 4, Zachary M. Mann offers a new way of seeing the Brobdingnag voyage in Jonathan Swift's *Gulliver's Travels.* Acknowledging the scholarly emphasis on the third voyage as a critique of Royal Society experimentalism and concerns about the production of scientific fact, Mann turns our attention instead to the changing industrial climate with which Swift was intimate as a political actor during the early eighteenth century. Mann argues not only that the textile industry and accompanying inventions of textile machines—the automated loom in particular—loomed large in Swift's imagination but also that *Gulliver's Travels* reflects Swift's awareness of and participation in the industrialization of intellectual labor.

In chapter 5, we shift from the loom to the steam engine, patented by James Watt in 1784, as Kevin MacDonnell considers an unexpected influence on the engine's design: William Hogarth's principle of the "line of beauty," which he presented in his 1753 *The Analysis of Beauty.* MacDonnell finds in particular that the most innovative component of Watt's 1784 engine—Watt's parallel linkage—"is born out of the same epistemological foundations that underpin" Hogarth's line of beauty. By examining the ostensibly superficial resemblance between Watt's linkage and Hogarth's line of beauty, MacDonnell demonstrates a common response, in the realms of technology and visual art, to "the geometrical straitjacket of contemporary design." MacDonnell extends this comparison of aesthetic principles to an analysis of Laurence Sterne's *Tristram Shandy* (1759–1767), showing how design principles of the historical Enlightenment not only contributed to what we now call Anthropocene ways of knowing and making but also undermined their implementation.

Chapter 6 invites a reconsideration of Horace Walpole's foundational Gothic novel *The Castle of Otranto* (1764) by offering a textual analysis of Walpole's "gothic lanthorn," which hung in the well of the staircase at Strawberry Hill, an estate that Walpole rebuilt in multiple stages in the style of the Gothic Revival. Walpole offered public tours of the house, and the lantern was among the first and most striking features visitors would see. Placing the lantern and its function within queer histories and histories of domesticity, Emily M. West shows how we can understand the lantern as an artifact of queer intimacies, arguing that the lantern presents a "critique of the heteropatriarchal family as a social form." In so doing, West brings the lantern as a technology into conversation with broader histories of optical technology in the period, demonstrating how Walpole's "curious" lantern is an artifact of queer materiality that challenged "linear narratives of rational modernity and the mechanisms conscripted to produce Enlightenment futures."

In chapter 7, Deven M. Parker turns our attention to an underexplored short story by Maria Edgeworth from 1807, "Lame Jervas," with the aim of investigating what the story reveals about eighteenth-century British anxieties about foreign technologies and the mechanisms used to alleviate them. The optical telegraph was, as Parker documents, a technology of particular concern around the turn of the century, due to France's recent telegraphic innovations and British fears about the possibility that it could be used as a tool of international surveillance. It was also, though, a promising technology for Britain, for it was perceived to afford new ways of protecting the nation's coastline from France. Maria Edgeworth was influenced by her father, Lovell Edgeworth, who had attempted (albeit unsuccessfully) to promote a telegraphic signaling device of his own invention, and her "Lame Jervas" expresses an optimistic view of telegraphic technologies, portraying them as a tool by which Britain may not only defend itself against foreign intrusion but also perform "British superiority" and, in so doing, position itself for pursuing its "imperial project." Thus, Parker's essay offers an example of one British literary author who harbored significant enthusiasm for the potential of technology to secure England's future and to foster progress as well as peace.

In chapter 8, Jamison Kantor pursues a provocative analysis of Percy Bysshe Shelley's poems *The Mask of Anarchy* (1819) and *The Triumph of Life* (1822) as guides for understanding the "political machine." Kantor argues that Shelley's interests in mechanical engineering and vehicular technology positioned him to offer a powerful critique of "state machinery" and the machine metaphor as a way of representing and governing social relations. In a departure from the critical focus on the notion of the finance economy as an automated, self-governing machine, Kantor reevaluates links between the political formation of the modern state and the development of industrial machinery.

While these chapters address a formidable range of texts, topics, and technologies, they have in common the aim of showing how even critical engagements with technology from roughly 1600 to 1830 helped shape technologies' functions and meanings in the world. By "shape," we mean that in some cases literary and aesthetic considerations contributed to the development of material technologies, while in others, the textual treatment of technology impacted how people understood and engaged with it. In other cases, as we have suggested, the literary text functioned as a technology in its own right, a tool that made something happen. In his 2017 monograph *Novel Machines: Technology and Narrative Form in Enlightenment Britain*, Joseph Drury emphasizes how profoundly the eighteenth-century novel registered the effects of technological innovation, "respond[ing] with new urgency to the concerns raised by Britain's so-called 'Industrial Enlightenment,' a culture of unprecedented technological innovation and ingenuity."[35] The present collection broadens this scope by surveying how literary works of various genres engaged with technological innovation. Like Drury, the authors here regularly show how literary works express "ambivalence" about the progress that was so often, and by so many people, associated with technological innovation.[36] We share Drury's "constructivist" recognition that "different machines prompt very different concerns in the eighteenth century, while the same machine may have different, even antithetical meanings in different circumstances."[37] By casting our net widely, considering not only a wide range of technologies but also a wide range of genres, we aim to expand scholars', teachers', and students' frameworks for understanding the multivalent relationship between literature and technology in the era of Enlightenment.

Much scholarly and public disputation over the legacies of the historical Enlightenment, particularly its scientific and technological legacies, has taken the form of opposition between those who understand the Enlightenment as a story of progress and those who are skeptical of progress narratives engendered by Enlightenment historiography. Although the chapters in this volume focus on texts and histories frequently critical of the progressive view of Enlightenment technology, they do so, as we have suggested, to provide a picture of the interrelation of literary text and material technology. We share Donna Haraway's view that "taking responsibility for the social relations of science and technology means refusing an anti-science metaphysics, a demonology of technology, and so means embracing the skillful task of reconstructing the boundaries of daily life, in partial connection with others, in communication with all of our parts."[38] Consequently, this volume functions less to give a thumbs-up or thumbs-down view of Enlightenment technological development than to provide a mosaic view of how literary texts checked and critiqued technological developments as well as helped constitute and enable such developments. In this mosaic view, writers of novels, poems, and plays do not stand athwart technological development yelling "Stop!" Rather, they reimagine the possibilities of technology and play a role in realizing those possibilities.

In recent decades, with the flourishing of environmental humanities, the Enlightenment has become a common target of blame for those who are seeking to identify the roots of the human tendency to cause environmental devastation. James Watt's patenting of the steam engine in 1784 is widely cited as the beginning of what has come to be known as the "Anthropocene," our current geological age in which human behavior has become the primary influence on our biosphere. According to this line of thinking, the Enlightenment is significantly responsible for having established the foundation for global warming and the mass extinctions with which we must now contend. Twenty-first-century ecocritics have sought to demystify much of what was perceived as "progress" by Enlightenment thinkers, countering the progressive view of the Enlightenment with narratives of destruction, exploitation, and degeneration. However, as the Enlightenment-era "Quarrel between the Ancients and the Moderns" attests, critiques of progress were far from uncommon in the Enlightenment era itself.[39] The present collection attests to how adept various literary authors of the Enlightenment were at scrutinizing technologies and the progress they so often were seen to represent. It also illustrates how common it was for them to do so. We therefore propose that the Enlightenment should be viewed as having laid the foundation not only for destructive conceptions of progress but also for the critique of such conceptions. The first steps toward interrogating Enlightenment technologies were taken from within, and literature was instrumental in paving the path.

In the early years of the nineteenth century, William Blake offered a skeptical portrayal of technology when, in his poem *Milton*, he asked, "And did the Countenance Divine, / Shine forth upon our clouded hills? / And was Jerusalem builded here, / Among these dark Satanic Mills?"[40] A distinguishing feature of the Industrial Age, technological innovation was inextricable from the development of modern factories (Blake's "dark Satanic Mills"). Thus, to critique the latter was to be at least circumspect about, if not equally critical of, the former. Such suspicion of technology persists today, though it has both expanded and become significantly qualified. One need not be a Marxist to bemoan the negative psychological impacts of "too much screen time," which has over the past decade or so been linked to depression.[41] However, technology has become so ubiquitous in the everyday life of inhabitants of the Global North that its presence tends to be taken for granted. Haraway argues that "by the late twentieth century, our time, a mythic time, we are all chimeras, theorized and fabricated hybrids of machine and organism—in short, cyborgs."[42] Inhabitants of the Global North have not only grown accustomed to technology but have internalized it to such an extent that it has become a part of our personal identities. How bereft most of us feel when rare circumstances require us to do without our cars, smartphones, and personal computers—not to mention our refrigerators and HVAC systems. When the power goes out in our homes, we hardly know what to do with ourselves. We may be suspicious of technology, but most of us cannot imagine living without it. Our volume registers the deep history of such ambivalence about technology. Writers during the historical

Enlightenment had not yet come to take the presence of technology for granted. However, they were, much like us, regularly both attracted to and repelled by its potential. Moreover, they had already begun to perceive themselves as "chimeras"—"hybrids of machine and organism." Our volume shows how literature across genres provided important sites for understanding and tinkering with this distinctively modern sense of the cyborgian self in the long Enlightenment period. Exploring such sites, we become better equipped to appreciate the modern self for what it is: "a creature of social reality as well as a creature of fiction."[43]

NOTES

1. Mary Ellen Bellanca, "Science, Animal Sympathy, and Anna Barbauld's 'The Mouse's Petition,'" *Eighteenth-Century Studies* 37, no. 1 (2003): 47–48.
2. Steven Shapin and Simon Schaffer, *Leviathan and the Air-Pump: Hobbes, Boyle, and the Experimental Life* (Princeton, NJ: Princeton University Press, 1985), 25.
3. Andrew Feenberg, *Questioning Technology* (London: Routledge, 1999), viii.
4. Neil Rhodes and Jonathan Sawday, "Introduction: Paperworlds: Imagining the Renaissance Computer," in *The Renaissance Computer: Knowledge Technology in the First Age of Print*, ed. Neil Rhodes and Jonathan Sawday (London: Routledge, 2000), 1–2.
5. Shapin and Schaffer, *Leviathan and the Air-Pump*, 25.
6. Ibid., 61.
7. Shapin and Schaffer chose the term "literary technology" with recognition of the etymology of "technology." They note Plato's division of *techne* into two forms, one associated with physical work and the other with speech. In this sense, "technology" already implies both the physical/mechanical and the verbal or literary. See *Leviathan and the Air-Pump*, 25n4.
8. See Joanna Picciotto, *Labors of Innocence in Early Modern England* (Cambridge, MA: Harvard University Press, 2010); Courtney Weiss Smith, *Empiricist Devotions: Science, Religion, and Poetry in Early Eighteenth-Century England* (Charlottesville: University of Virginia Press, 2016); Joseph Drury, *Novel Machines: Technology and Narrative Form in Enlightenment Britain* (Oxford: Oxford University Press, 2017); Al Coppola, *The Theater of Experiment: Staging Natural Philosophy in Eighteenth-Century Britain* (Oxford: Oxford University Press, 2016); and Tita Chico, *The Experimental Imagination: Literary Knowledge and Science in the British Enlightenment* (Stanford, CA: Stanford University Press, 2018).
9. Feenberg, *Questioning Technology*, 7.
10. Ibid., 11.
11. Ibid., 14, vii.
12. Jennifer L. Lieberman, "Finding a Place for Technology," *Journal of Literature and Science* 10, no. 1 (2017): 26.
13. Ibid., 27.
14. D.S.L. Cardwell, "Science, Technology, and Industry," in *The Ferment of Knowledge: Studies in the Historiography of Eighteenth-Century Science*, ed. Roy Porter and G. S. Rousseau (Cambridge: Cambridge University Press, 1980), 452.
15. Ibid., 450.
16. Paul Forman, "The Primacy of Science in Modernity, of Technology in Postmodernity, and of Ideology in the History of Technology," *History and Technology* 23, nos. 1–2 (2007): 53.
17. Robert Hooke, *Micrographia; or, Some Physiological Descriptions of Minute Bodies Made by Magnifying Glasses* (London: Martyn and Allestree, 1665), preface.
18. Lieberman, "Finding a Place for Technology," 25.
19. David Spadafora, *The Idea of Progress in Eighteenth-Century Britain* (New Haven, CT: Yale University Press, 1990), 76–84. See also Roger Hahn, "The Application of Science to Society: The Societies of Arts," *Studies on Voltaire and the Eighteenth Century* 25 (1963): 829–836; James E. McClellan III, *Science Reorganized: Scientific Societies in the Eighteenth Century* (New York: Columbia University Press, 1985).

20. Francis Bacon, *The New Organon*, trans. and ed. Lisa Jardine and Michael Silverthorne (Cambridge: Cambridge University Press, 2000), 13.

21. Ibid., 91.

22. Spadafora, *The Idea of Progress*, 84.

23. Lisa Jardine, introduction to Bacon, *The New Organon*, xiv.

24. William Shipley, "A Proposal to Establish a Society for Promoting Useful Knowledge in the County of Kent. Addressed to the Inhabitants of the Said County" (1786), printed in D.G.C. Allan, *William Shipley, Founder of the Royal Society of Arts: A Biography with Documents*, corr. ed. (London: Scolar Press, 1979), 203.

25. Spadafora, *The Idea of Progress*, 76–84.

26. Gulliver's Brobdingnag voyage is often read as an allegory of modern microscopy. See, for instance, Kristin M. Girten, "Mingling with Matter: Tactile Microscopy and the Philosophic Mind in Brobdingnag and Beyond," *The Eighteenth Century* 54, no. 4 (2013): 497–520; Deborah Needleman Armintor, "The Sexual Politics of Microscopy in Brobdingnag," *SEL* 47, no. 3 (2007): 619–640; Christopher Fox, "Introduction: How to Prepare a Noble Savage: The Spectacle of Human Science," in *Inventing Human Science: Eighteenth-Century Domains*, ed. Roy Porter Fox and Robert Wokler (Berkeley: University of California Press,1995), 1–30; Marjorie Nicolson, "The Microscope and English Imagination," *Smith College Studies in Modern Languages* 16, no. 4 (1935): 1–92.

27. Jonathan Swift, *Gulliver's Travels*, in *The Prose Works of Jonathan Swift*, vol. 11, ed. Herbert Davis (Oxford: Basil Blackwell, 1941), 118.

28. Ibid., 118–119.

29. Ibid., 119.

30. Margaret Cavendish, *Observations upon Experimental Philosophy*, ed. Eileen O'Neill (Cambridge: Cambridge University Press, 2001), 51.

31. Ibid., 50–51.

32. Ibid., 51.

33. Ibid., 52.

34. Jonathan Swift, "A Discourse Concerning the Mechanical Operation of the Spirit," in *The Essential Writings of Jonathan Swift*, ed. Claude Rawson and Ian Higgins (New York: Norton, 2010), 125.

35. Drury, *Novel Machines*, 2.

36. Ibid.

37. Ibid., 10. For the precise sense of "constructivism" adopted by Drury, see Feenberg, *Questioning Technology.*

38. Donna Haraway, "Manifesto for Cyborgs: Science, Technology, and Socialist Feminism in the 1980s," in *Simians, Cyborgs, and Women: The Reinvention of Nature* (New York: Routledge, 1991), 149. This essay was originally published in *Socialist Review* 80 (1985): 65–108.

39. According to Fredric Jameson, this Quarrel helped to usher in the "birth of historicity." Jameson, *A Singular Modernity: Essay on the Ontology of the Present* (London: Verso, 2013), 21–22.

40. William Blake, *Milton*, in *The Complete Illuminated Books* (New York: Thames and Hudson, 2000), 446.

41. See, for instance, Elroy Boers, Mohammad H. Afzali, Nicola Newton, and Patricia Conrod, "Association of Screen Time and Depression in Adolescence," *JAMA Pediatrics* 173, no. 9 (2019): 853–859.

42. Haraway, "Manifesto for Cyborgs," 149.

43. Donna Haraway, "A Cyborg Manifesto: Science, Technology, and Socialist-Feminism in the Late Twentieth Century," in *Simians, Cyborgs, and Women*, 149.

CHAPTER 1

Webster's Baroque Experiments and the Testing of Technology in the Early 1600s

Laura Francis

While the latter half of the 1600s may have witnessed England's most famous contributions to the Scientific Revolution, Jacobean London was already teeming with the technological activity that would help usher it in. Clockmakers, lens grinders, and mathematicians settled across several neighborhoods.[1] Mechanisms made noise throughout the streets.[2] Advertisements for the newest devices competed with playbills for attention around the Blackfriars' district.[3] Technology was a part of daily life as always, and the latest innovations were often spurred by cross-cultural exchanges at the time: as the scholars Deborah Harkness and Adam Cohen have shown, the translation of numerous technological treatises as well as the immigration of Dutch, Italian, French, and Spanish instrument makers created networks of influence that reached well beyond the borders of the burgeoning English empire.[4] Galileo, Kepler, and others may have been mapping the heavens with their refracting telescopes from the continent, but their technologies—as well as the consequences of their observations—were making their way into London.

This chapter explores how contemporary writers helped shape such exchanges. According to Michel Foucault's seminal, if debated, arguments in *The Order of Things*, the influx of technology also came with a wave of epistemological shifts and representational problems. Sixteenth-century Europe relied heavily on linguistic relations like *aemulation*, *convenientia*, *analogy*, and *sympathy/antipathy* to structure humanity's relationship to nature and the heavens, but a tabular order of things began to overtake this system in the Enlightenment: *mathesis*, *taxinomia*, and *genesis* began to define relationships on empirical data rather than the metaphorical "prose of the world."[5] Such large-scale changes in the organization of knowledge do not occur in a day, of course, and historians of science have already shown how natural philosophers were working through the epistemic problems coming hand in hand with the development of technology.[6] Ofer Gal and Raz Chen-Morris have argued that the unprecedented powers being ascribed to instruments created an alienating dilemma throughout the 1600s, for example. They assert that the idea of granting authority to devices meant admitting that the "human eye is nothing but an instrument, and a weak one at that. . . . Paradoxically, accurate scientific observation and the naturalized understanding of the senses detached the intellect from its objects."[7] Empirical precision thus came with disturbing implications for the interconnected nature of body and mind. The very language of truth was moreover at stake as intellectuals like Sir Francis Bacon debated whether figurative or "plain" speech could better communicate the evidence of observation to others.[8]

It is perhaps unsurprising that the representational problems surrounding technology and observation involved artists as well as natural philosophers, then. Historians have also shown how painters, particularly baroque painters, participated in such epistemic changes. Gal and Chen-Morris argue that the extreme character of the baroque can be seen as more than a decadent extroversion or illogical playfulness amid the rise of the tabular orders. It is "a genuine investigation of the possibilities of the artificial sense organs," a counterpart rather than anathema to the "mathematical New Science," a study of "orderliness and the problematic means of producing it."[9] The likes of Rubens, Bettini, and Vermeer thus questioned the changing organizations of knowledge in their works on sensory perception alongside natural philosophers of their day.[10] Pushing this even further, recent scholarship has proposed that paintings, drawings, and engravings were instruments for the artist and astronomer alike. Both used such tools to shape their arguments for distant readers, and both made vision itself into a performance.[11] The demonstrations of Brunelleschi's mathematical perspective and Galileo's refracting telescope similarly turned the various *ways* of viewing the world into spectacles for viewers to judge.[12]

English literature's contributions to such dilemmas have not been studied quite as fully in this active sense, however. While great strides have been made to understand the presence of technology in writing, such as foundational studies on technological metaphors in Shakespeare, they tend to emphasize literature's role as receptor, consumer, and/or mediator for public opinion rather than potential contributor itself.[13] This unidirectional trend leaves much room for exploration in the history of intersections between the arts and sciences in a time when they were not necessarily considered separate fields of knowledge.[14] How did authors respond to the alienating effects of technological advances and movements into the tabular orders, then? And, like baroque painting and mathematical perspective, how might we see writing as an instrument that turned these changing ways of seeing the world into spectacles for audiences to judge? Following the work of art historians, as well as recent studies pushing our understanding of literary engagement with the sciences, I posit that English literature of the early 1600s not only demonstrated an awareness of contemporary technology and natural philosophy but also explored their boundaries.[15] Poets and especially dramatists turned their contemporaries' tools and theories into performances to test the extent of their truth-making powers while challenging audiences to do the same.

John Webster's *The Duchess of Malfi* (performed 1613; published 1623) offers an ideal case study for investigating this kind of experimental engagement. Although the relationship between Webster's tragedy and contemporary anatomy theaters has been explored quite extensively, few studies have analyzed its relationship to technology or natural philosophy.[16] Scholars have traditionally tended to interpret the play in relation to tainted bloodlines, the rise of the merchant class, and moral philosophy, all indeed compelling lines of criticism considering the primary subject matter.[17] Two of the main characters (Ferdinand and the Cardinal) punish their sister (the Duchess) for marrying a man below her rank without their consent, and the result is a bloody spectacle rife with strange plots, wax figures of her lifeless family, the death of almost every character, and the establishment of a tenuous new line through her half-aristocratic, half-bourgeois son. Yet the play also contains a rich engagement with the sciences of the day. Webster uses over thirty metaphors involving contemporary instruments and mathematics, regularly discusses the effects of the senses on the mind, and employs a destitute university graduate as its primary intelligencer-artificer, all providing rich material for understanding how playwriting could work as a testing ground for the changing order of things.

If we attend to the representation of technology and natural philosophy in Webster's work, we can see how he turned his contemporaries' devices and ways of seeing into spectacles for his audiences to judge. His speeches load the characters' interactions with technological references. His torture scenes reveal how his wax puppets physically embody the classical metaphors for sensory perception to test the precision of bodily instruments. His strange echoes emphasize how the script plagues spectators as much as characters in its sensory dilemmas. In sum, *The Duchess of Malfi* demonstrates how playwrighting can be seen as an instrument not in the sense of absolutist propaganda but rather in the sense of experiment. Considering the contemporary definitions for "instrument" ("a device or apparatus for registering, measuring, or recording a physical quantity, property, or phenomenon"), "experience" ("tentative procedure; an operation performed in order to ascertain or illustrate some truth; an experiment"), and "experiment" itself ("an action or operation undertaken in order to discover something unknown, to test a hypothesis, or establish or illustrate some known truth"), the play serves as a particularly baroque testing ground for the capacities of developing devices and natural philosophy alike.[18] It provides a set of controls (e.g., script) that test variables (e.g., audience) as it measures the effects of technology and the tabular orders in a range of situations. Like prior work on the artist's brush, it might serve as a model for understanding active literary engagement with the sciences of the time.

Wax Metaphors and the Instruments of Vision

Webster's baroque experimentation can be seen from the most fundamental level of vocabulary to the most conceptual level of philosophy throughout the play. Beginning with the most rudimentary of elements, in a way, the dialogue consistently references contemporary devices, natural philosophy, and mathematics. In relation to technology, the characters utilize words like the following: spectacles, hour-glass, cupping-glass, ventage, anvil, forge, hammer, pot-gun, musket, wire-drawer, enginous wheels, hinges, clock (2), watch (2), instrument (2), gun-powder (2), mine (2), cannon (3), compass (3), engine (3), glass-house (3), bullets (4), pistols (4), and glass (7). In relation to astronomy and mathematics, they employ language like the following: circumference, base, pyramid, point, multiply, spheres, halves, lines, mathematics, mathematical, geometrical, geometry (2), perspective (2), eclipse (2), remainder (2), motion (3), divide (4), form (6), shape (6), none (16), and nothing (24). In relation to sensory perception and the sciences more generally speaking, they bring in vocabulary like the following: experiment, almanac, dictionary, hearing, physiognomy, eyelid (2), eye-sight (2), tune (2), physic (3), examine (3), ears (7), echoes (8), study (10), physician (10), noise (11), hear (22), eyes (25), and see (48). The characters' interactions are thus laden with the language of scientific knowledge at the time.

Such frequencies may demonstrate the simple fact that new technologies were becoming a part of everyday life for audiences at the Blackfriars' Theatre. The dialogue is certainly brimming with references in a range of contexts throughout the play: characters jest that they would like to have a "mathematical instrument" made for a woman's face to keep her from laughing "out of compass"; claim that the Duchess will marry before the "turning of an hourglass"; and chide servants for their ignorance of math, claiming that the devil likes to "hang at a woman's girdle, like a false rusty watch," so that she cannot "discern how the time passes."[19] Even Galileo's "fantastic glass" appears in the dialogue with the hope of viewing "another spacious world . . . to find a constant woman there."[20] The consistency could indeed mark the growing place of such instruments in contemporary life.

Yet the references begin to reveal something beyond the everyday. Webster does not just employ them in simple figures of speech but often renders them absurd by their own hyperboles. The instruments cannot possibly measure the spatial and temporal elements claimed by the characters: Galileo's glass cannot extend its view to an otherworldly region of chaste women, nor can an hourglass determine time until an unknown marriage, nor can a mathematical device hold a lady's face in check. The scale is off. The instruments are twisted into the service of an unsettling control over others' bodies as the playwright aligns much more closely with his own madman than someone invested in the powers of new technology: "If I had my glass here," his character brags, "I would show a sight should make all the women here call me a mad doctor."[21] Webster's theatrical lens does potentially show a sight that could make his own audience call him mad. His obsession with hyperboles that delight in their own absurdity reveals a strange testing ground for technology and the language of truth in the day. The characters quickly face unreadable surfaces and even more unreadable depths as the metaphorical instruments remain useless to understand or alter their bloody fates.

Indeed, the dialogue grows increasingly obsessed with the inability to trust one's eyes, to see below the surfaces of the world, to discern false appearances from true substances—all particularly baroque and scientific concerns of the time.[22] Even as advancing technologies allowed natural philosophers to observe the smallest and widest of worlds with ever-refining microscopes and telescopes, their developments produced as many questions as answers. As summarized by Édouard Glissant, the sciences of the 1600s "postulated that reality could not be defined on the basis of its appearance, that it was necessary to penetrate into its 'depths,' but it also agreed that knowledge of these was always deferred, that no longer were there grounds for claiming to discover the essentials all of a piece." This led to an age of rational uncertainties and the "very basis of the baroque tendency" that would define future worldviews.[23] Thus, following Gal and Chen-Morris's argument, the strangeness of the baroque becomes a study of orderliness, a genuine investigation of the possibilities brought by the artificial sense organs and the fundamental estrangement from reality that came with them.[24] Webster's work toys with not only the unsettling use of devices like "compass" and "glass," then, but also the bodily instruments they attempt to enhance, like one's eyes and even the human mind itself.

The play begins probing nature's "depths" and the alienation of bodily instruments through the baroque trope of appearances that do not match substances, consistently displaying Glissant's argument that there were no longer "grounds for claiming to discover the essentials all of a piece." When a character asks about the Cardinal, for instance, the Duchess's husband, Antonio, replies that some flashes of bravery "superficially hang on him for form; but observe his inward character." He is "nothing but the engendering of toads," strewing "flatterers, panders, intelligencers, atheists, and a thousand such political monsters" in his way.[25] His brother, Ferdinand, elicits similar readings: "The Duke there?" Antonio says. "A most perverse and turbulent nature. Whatever seems in him mirth is merely outside."[26] Their appearances clearly do not match their substances, and the characters express their disgust for such disjunctions in terms of monstrosity, loose skin, and slime. Their sister does not escape from these kinds of readings, either. Antonio wishes that other ladies would "dress themselves in her."[27] Ferdinand asks what "hideous thing" eclipses her virtue.[28] The Cardinal complains that she makes "religion her riding hood / To keep her from the sun and tempest."[29] The Duchess's beauty and holiness thus appear like cloaks or detached skins, and even her pregnancy is represented as a disjunction between her aristocratic exterior and mixed-class child. While these disconcerting lines initially focus on the siblings—perhaps a commentary on tainted structures of aristocracy—the whole play becomes increasingly consumed by the inability to trust one's perception of exteriors or to reconcile them with their knowledge of interiors.

The Duchess's servant Cariola exclaims that her innocence will be seen on her heart when she is murdered, for example, an anxious assertion that others will perceive her appearances incorrectly.[30] The intelligencer Bosola similarly admonishes his "painted honor" and asks another character, "What thing is in this outward form of man / to be beloved?" before deriding humanity's meaningless representations on earth:

> But in our own flesh, though we bear diseases
> Which have their true names only ta'en from beasts,
> As the most ulcerous wolf and swinish measle,
> Though we are eaten up of lice and worms,
> And though continually we bear about us
> A rotten and dead body, we delight
> To hide it in rich tissue: all our fear,
> Nay all our terror is lest our physician
> Should put us in the ground, to be made sweet.[31]

Instead of a house for the soul, the body is already "rotten and dead." It is a senseless artifice to be eaten by bugs, and even its infections' "true names" are taken from beasts rather than humans. Ultimately, the "rich tissue" of both dress and flesh is an exterior that people wear in order to hide their terror. It is sumptuous, the dialogue relishing in its materiality rather than transcendence or careful observation.

When Ferdinand presents the Duchess a wax version of her murdered family, the inability to trust one's perception of exteriors or probe interiors is then distorted to baroque extremes. The false substance is so convincing that it overrides both her sensory perception and judgment, and the ambiguous dialogue leaves even spectators to wonder what is real within the play-world. First, Ferdinand supposedly gives the Duchess her husband's hand:

> FERDINAND It had been well
> Could you have lived thus always, for indeed
> You were too much i'th'light. But no more;
> I come to seal my peace with you. Here's a hand
> *Gives her a dead man's hand.*
> To which you have vowed much love; the ring upon 't
> You gave.
> DUCHESS I affectionately kiss it.
> FERDINAND Pray, do, and bury the print of it in your heart.[32]

The Duchess does not balk at the sight of the hand or claim that it is fake. Though the reaction could vary depending on the actor's choices in gestures, the dialogue suggests that she kisses it in belief. When Ferdinand then asks her to "bury the print of it" in her heart, he is moreover asking her to imprint this representation into her body as if it were real: the appearance is fully estranged from the true substance, yet the empty representation is taken for Antonio's real, fleshy limb. Even accounting for lighting and prop choices, this could also force audience members to use their judgment to override their initial perception and distinguish the real from the false.[33]

The dilemma only becomes more twisted and complex as the scene develops. When Ferdinand leaves the Duchess with the fake hand, he opens up the curtain to an inner scene as wax and flesh, instrument and body, subject and object, all become almost impossible to distinguish from one another:

> FERDINAND Let her have lights enough. [*Exit.*]
> DUCHESS What witchcraft doth he practice, that he hath left
> A dead man's hand here?
> *Here is discovered, behind a traverse, the artificial figures of* ANTONIO *and his Children, appearing as if they were dead.*
> BOSOLA Look you, here's the piece from which 'twas ta'en.
> He doth present you this sad spectacle,
> That, now you know directly they are dead,
> Hereafter you may wisely cease to grieve
> For that which cannot be recoverèd.
> DUCHESS There is not between heaven and earth one wish
> I stay for after this; it wastes me more
> Than were't my picture, fashioned out of wax,
> Stuck with a magical needle, and then buried
> In some foul dunghill; and yond's an excellent property.
> For a tyrant, which I would account mercy
> . . . .
> Let Heaven a little while cease crowning martyrs
> To punish them!
> Go, howl them this, and say, I long to bleed:
> *It is some mercy when men kill with speed.*[34]

Only the stage directions let on that the figures of Antonio and his children are "artificial," and the Duchess takes this "sad spectacle" as proof that they are dead and unrecoverable even as she exclaims that it wastes her more than a tortured picture "fashioned out of wax." For the audience watching the play-within-the-play, Webster does not give any clues about the corpses for almost two pages of dialogue either. As scholars have asserted, then, these tricks could have easily fooled the perception of original audiences along with the Duchess within the play: while the King's men might have hired an artist to create a mannequin for the "artificial figures of Antonio and his children," it is likely that the actual actors would have played these dead bodies and therefore confounded the boundaries between artificial figures and actual flesh.[35]

Whether skin or wax, spectators inside and outside the play are thus left suspended in the illusion. When Webster does finally reveal the trick, it comes with Ferdinand's boast about the success of his experiment:

> FERDINAND Excellent, as I would wish; she's plagued in art:
> These presentations are but framed in wax
> By the curious master in that quality,
> Vincentio Lauriola, and she takes them
> for true substantial bodies.[36]

His trick is only "framed in wax," but he delights in how the Duchess takes the horrifying depiction of her dead loved ones to be "true substantial bodies." He is excited that she is "plagued in art" by the "curious master" who made the spectacle possible. The reference to "Vincentio Lauriola" has not been substantiated with any historical connections, but it moreover signals an appreciation for the technical precision of waxwork.

Indeed, wax was known for its value across a range of disciplines at the time. As Lynn Maxwell describes in *Wax Impression, Figures, and Forms in Early Modern Literature*, its moist and "palpable" qualities made it an ideal substance for the work of artists, anatomists, and philosophers alike. Sculptors used it in order to work out conceptual problems before committing to their final material.[37] Painters used it to create "perfectly manipulable subjects" that they could control in ways that were simply not possible with human models.[38] Anatomists favored it because of its cleanliness, especially its ability to represent organs such as the kidneys, eyes, and even unborn fetuses persuasively for female viewers without the blood of anatomical demonstrations.[39] Later, in the 1700s, wax museums would even be supported by the future pope Prospero Lambertini, while writers used the substance for their own conceptual problems in the form of erasable tablets.[40] Philosophers like Cicero, Aristotle, Descartes, Locke, and others similarly favored it in their figurative representations of the mind, sensory perception, and judgment.[41]

It is this last grouping of natural philosophers that most concerns us here. The "curious master" in *The Duchess of Malfi* brings questions of sight and judgment to the forefront as the fall of the curtain on the "sad spectacle" does little to mitigate the confusion produced by the artificial figures. When the Duchess asks Cariola how she seems after the incident, the servant tells her mistress that she looks like a spectacle of appearance rather than substantive flesh:

> Like to your picture in the gallery:
> A deal of life in show, but none in practice;
> Or rather, like some reverend monument
> Whose ruins are even pitied.[42]

The Duchess, too, seems to have been inverted by the experiment. Substance has become appearance. Body has become instrument. Subject has become object. As Maxwell argues in her chapter on *The Duchess of Malfi*, the picture is as much like the woman as the wax figures are like her family.[43] The Duchess moreover replies that this inversion is quite proper, for "Fortune seems only to have her eye-sight / to behold my tragedy."[44] Sight is thus presented as a kind of possession for Fortune. It is an instrument that can do little more than observe, that can only look upon tragedy, and even then in a twisted fashion. Though the play certainly deals in ideas of "wax magic" and a picture "fashioned out of wax," according to Maxwell, I would add that Ferdinand's illusion makes noteworthy associations with the ideas of sensory perception among natural philosophers. Could we not see Webster's playwriting as an engagement with the sciences as well? Could we not see the wax as a tool in his own kind of baroque thought experiment?

When the play is considered in the context of treatises on perception and the body's sensory instruments, it becomes clear how Webster takes the philosophers' conceptual problems and fashions their ways of seeing the world into his strange puppets. As Gal and Chen-Morris explain, Aristotle defines "sense" as "what has the power of receiving into itself the sensible forms of things without the matter," just as a "piece of wax takes on the impress of a signet-ring without the iron or gold."[45] For him, the wax metaphor emphasizes the direct contact of the object with the sense organ through a medium, reinforcing the unity of the senses and the idea that remote objects of vision are represented as reliably as the immediate objects of touch. According to Jacques Lezra, Renaissance commentators also took up this metaphor as their model and continued to organize concepts of the body and soul through it. When Descartes uses it in the seventeenth century, then, it comes as no accident.[46] He draws on and departs from a long lineage of wax: "Sense perception occurs in the same way that wax takes an impression from a seal. . . . We must think of the external shape of the sentient body as being really changed by the object in exactly the same way that the shape of the surface of the wax is altered by the seal. . . . Thus, in the eye, the first opaque membrane receives the shape impressed upon it by multi-coloured light."[47] Unlike Aristotle, Descartes de-emphasizes what touches the "opaque membrane" here. It is not the properties of the visible, corporeal body but rather the multicolored light that makes the impression, and the same is true for all the senses.[48] Locke would also use wax to describe the human sensory system and perception even decades later: "If the Organs, or Faculties of Perception, like Wax over-hardned with Cold, will not receive the Impression of the Seal, from the usual impulse wont to imprint it; or, like Wax of a temper too soft, will not hold it well, when well imprinted; or else supposing the Wax of a temper fit, but the Seal not applied with a sufficient force, to make a clear Impression: In any of these cases, the print left by the Seal, will be *obscure*."[49] Again, the contact between "wax" and "seal" is used as an analogy for perception. If the sensory organs are hardened like frozen wax, they "will not receive the Impression of the Seal, from the usual impulse wont to imprint it"; if they are overly sensitive like warm wax, they "will not hold it well, when well imprinted" either.

When Ferdinand comes to "seal his peace" with the Duchess and lets "her have lights enough," then, I would argue that the dialogue uses an exaggerated, embodied form of the wax metaphor that breaks with Aristotle. Webster's scene displays an inharmonious impression of the performance on the character's senses: Ferdinand's figures are real enough to "imprint" on the Duchess's faculties of perception but, as Descartes later asserts, are not exactly a "direct representation" of her family either. The bodily instruments find their limit in the experiment, while the numerous technologies referenced throughout the play provide no assistance.

According to the philosophers, judgment could provide some hope here. Descartes later emphasizes the importance of the mind to observe the world around us despite our faulty senses, for example. He first observes the properties of cold wax in his meditations: easy to touch and still fragrant and flavorful with the light taste of honey, in sum, all the properties that "seem to be required for a given body to be known as distinctly as possible." Yet he also imagines them as they melt away by the fire: all of those qualities discovered through taste, smell, sight, touch, and hearing have changed. In the midst of such alterations, only judgment can determine that this shifting substance is still wax. The "flavor of its honey," the "scent of the flowers," and the "cold" texture evaporate and melt before the senses, but only the intellect can say the "wax remains."[50] Such exercises ultimately lead Descartes to the idea of *cogito*, even though God alone can provide the ultimate guarantee against uncertainty and even madness. Diverging from this thought experiment, Locke purports that the senses are the primary source of human understanding and that other bodies contain essential qualities within themselves.[51] Wax does possess inherent characteristics that produce the sensations of color and sound in the sensory organs, operating on them and other bodies just as the sun shines on wax and melts it. These operations are moreover what give the human mind an "idea" of such "artificial things." Judgment then takes into account the mind's experience of other shapes to assess this information.[52]

When the Duchess takes Ferdinand's "artificial" figures for "true substantial bodies," I would argue that the play shows how both the senses and judgment can be deceived by the likeness of their qualities to her real children. The palpable wax fools the spectators within the play-world, while the actors' bodies potentially fool the spectators observing from their places in the crowd outside it. Webster thus displays how the Duchess's and audience's prior knowledge of her family's shapes informs the judgment of the sensory input, though the erroneous conclusion suggests a skeptical view on the corroboration of evidence. While God might have still provided some certainty in perception and/or judgment for philosophers of the 1600s, Webster's wax offers no such comfort.[53]

As Rainer Nägele explains in *Theater, Theory, Speculation: Walter Benjamin and the Scenes of Modernity*, baroque tragedy is a "radical secularization of the medieval mystery plays that empties out all transcendence and leads toward a cancellation of all transcendence."[54] The body tends to oscillate between "physical entity" and "imaginary subject/object" in this voiding process, and it ultimately focuses on the sheer corporeality of the body: the drama thus confronts "the audience with the unmetaphorical interior, literally with the guts and 'innards' of the opened body."[55] The exterior is radicalized to the point that the actors are replaced by false puppet figures. The surface, the artifice, the show—they become indistinguishable from substantive flesh like the instrumentalized, alienated body of the scientific observer, and neither judgment nor perception nor assurance from God can distinguish them. *The Duchess of Malfi* therefore uses a particularly baroque, particularly skeptical embodiment of the wax metaphor to experiment with the instruments of the human body and their capacity to probe the depths of the world.

Experimental Echoes

The sensory dilemmas do not end after the use of wax or even the Duchess's death, though, especially for the audience. While Ferdinand's puppets may disturb observers offstage in act 4, Maxwell remarks that spectators should always know or at least assume that the action is fictional even if they do not understand the exact nature of the puppet. It is a play, after all. Yet, she asks, do they?[56] I would argue that Webster refuses the audience any easy assumptions or mental distance from the stage as the plot thickens. Beyond the strange implementation of contemporary devices in the dialogue, beyond the radical embodiment of wax metaphors in natural philosophy, the final act reaches out to test the audience through sound.

Early on in the play, Ferdinand hopes that his sister and her family are "burnt in a coal pit with the ventage stopped, / that their cursed smoke might not ascend to heaven."[57] Indeed, his wish is granted when her voice remains behind after he hires Bosola to kill her. When Antonio and Delio enter the stage in the final act, an "*Echo from the* DUCHESS' *grave*" enters with them:

> ANTONIO And, question-less, here in this open court,
> Which lies naked to the injuries
> Of stormy weather, some men lie interred
> Loved the church so well, and gave so largely to 't,
> They thought it should have canopied their bones
> Till doomsday; but all things have their end:
> Churches and cities, which have diseases like to men,
> Must have like death that we have.
> ECHO *Like death that we have.*
> DELIO Now the echo hath caught you.
> ANTONIO It groaned, methought, and gave
> A very deadly accent.
> ECHO *Deadly accent*
> . . . .
> ANTONIO 'Tis very like my wife's voice.
> ECHO *Ay, wife's voice.*[58]

This is a space bereft of transcendence and substance in true baroque fashion. The court now lies open to the earthly elements, with holy buildings and men exposed to an end that is not the "doomsday" they expected: churches and cities will have a senseless and diseased death without redemption, and the remainders of a deteriorated body echo in a "deadly accent" over the graves. The scene also tests the audience's sense of hearing to determine whether this "deadly accent" belongs to Antonio, the Duchess herself, or some other entity. Up to this point, audience members could maintain a privileged perspective over the stage; here, however, they are left blind to contemplate the environment before them.[59] Unlike the dead hand or wax figures, distanced from the audience, the echoes envelop them in their sound and leave them "plagued in art" as well.

The echo should not be passed over as just any instrument of terror, either. Following Walter Benjamin's seminal study on baroque tragedy, the antithesis of coherent speech occurs when sound and meaning are brought together but do not form an organic linguistic structure. The echo, for example, repeats the last syllables of a strophe and thus conforms to the logic of the original line; yet its sounds are detached from the original and can assert their own meanings. In other words, the second half could form a new phrase with new significance. If the word "underlying" were to echo in a room, for example, the sound might result in something like "underlying," "-lying," "-lying," "-lying," and so on. The word for describing a location thus transforms into an indication of dishonesty with the repetition of "lying" over and over again. *The Duchess of Malfi*'s echoes certainly follow this line of thought. They repeat the logic of Antonio's sentences (in the structure of "like death that we have," for instance), but they also become active forces of their own ("the echo hath caught you," as Delio remarks).

Benjamin takes his argument on baroque sound even further than this, though. When speech is re-echoed from the mouth of the intriguer and "master of meanings," he argues that it creates a violent distortion that comments on the nature of language itself: "If the echo, the true domain of the free play of sound, is now, so to speak, taken over by meaning, then it must prove to be entirely a manifestation of the linguistic, as the age understood it."[60] This is a bold assertion, but the tension between logic and meaning in Webster's play may indeed speak to the relationship between period and language. In the context of the hyperbolic references to instruments, the failure of both sense and judgment, and the death of the Duchess at the hands of the university-trained intelligencer Bosola, I suggest that the graveyard echoes express profound skepticism about the viability of a world controlled in any extreme tabular order of things. The linguistic remainders here emphasize the "death" and "deadly" accents reflecting and refracting amid the ruins of the play.

I also suggest that the echo "taken over by meaning" reveals an important aspect of playwriting and of the theater as instrument. The scene stages the echo as part repetition, part agent. Unlike Benjamin's example, where the intriguer uses the echo to actively toy with his victim, the echo in Webster's play reverberates from the victim's grave:

> DELIO Come, let's walk farther from 't
> I would not have you go to th' Cardinal's tonight—
> Do not.
> ECHO *Do not.*
> DELIO Wisdom doth not more moderate wasting sorrow
> Than time: take time for 't—be mindful of thy safety.
> ECHO *Be mindful of thy safety.*
> ANTONIO Necessity compels me.
> Make scrutiny through the passages of your own life; you'll find it impossible
> To fly your fate.
> ECHO *Oh, fly your fate!*[61]

In this case, the echo attempts to advise Antonio. It attempts to avert further bloodshed by emphasizing Delio's recommendations to "Be mindful of thy safety" and "Oh, fly your fate!" This inversion of Benjamin's example allows both characters and audiences to re-inscribe the sound with significance. Unlike the disappearing observer of the new sciences, with their sensory organs alienated from themselves, Webster surrounds his viewers in the sound. The play leaves the freed, fragmented representation for spectators to contemplate and fill, to consider the possibilities and limits of their powers to define as the voices resound through the theater. The script may provide a control for Webster's experiments, but it leaves the experience of each audience member open as variables to measure.

The last lines question whether the impression of the performance will take, for instance, emphasizing how the sun "melts both form and matter" of the events like the wax of Descartes and Locke:

> DELIO . . . Let us make noble use
> Of this great ruin; and join all our force
> To establish this young hopeful gentleman
> In 's mother's right. These wretched eminent things
> Leave no more fame behind 'em, than should one
> Fall in a frost, and leave his print in snow;
> As soon as the sun shines, it ever melts,
> Both form and matter. I have ever thought
> Nature doth nothing so great for great men
> As when she's pleas'd to make them lords of truth:
> *Integrity of life is fame's best friend,*
> *Which nobly, beyond death, shall crown the end.*[62]

The urge to make "noble use of this great ruin" could be seen as an ironic comment on the play's violence, considering that the impression of the bloody events might not easily disappear to make room for the "integrity of life." It might also ironically question whether Nature truly "doth nothing so great for great men / as when she's pleas'd to make them lords of truth." The rest of the play offers embodiments that probe the trustworthiness of human senses and judgment and potentially critiques any extreme shift toward the tabular orders that crown humans as the "lords of truth" by rendering them as absurd as the hyperbolic references to their instruments. Rather than crown any one master of nature, the play leaves audience members—however faulty their judgment or senses—to remold the representations with their own interpretations after the final lines melt away. One can only hope, perhaps, that integrity and humility will be humanity's "friend" as the world advances beyond the play's end.

CONCLUSION

When considered in the context of contemporary sciences, *The Duchess of Malfi* presents not only an awareness of technology and natural philosophy but also an active response to them on various levels. Its dialogues brim with references to technology; its wax embodies the metaphors of natural philosophy to test the bodily instruments of perception and judgment; and its echoes test the audience's ability to distinguish the artificial from the real, demonstrating how the theater can become an instrument registering the changing orders of the world as the playwright experiments with subject and object alike. In sum, like the visual arts of the time, the play puts various ways of seeing and understanding the world on baroque display for its spectators to judge.

Webster was not alone in such an endeavor, either. We might see other writers of the early 1600s as commentators on natural philosophy, their metaphors not just sensible tools for reading the world around them but active distortions, questions, and contributions to their scientific contemporaries. The uses of hyperbole in John Donne's *Anniversaries* could be seen as stretching the limits of anatomy's potential to produce truth; the constant "nothings" of Shakespeare's *King Lear* could be seen as testing the power of mathematics to know what it means to have and be "zero"; the profusions of Robert Burton's *Anatomy of Melancholy* could be seen as pushing the limits and possibilities of encyclopedic knowledge. When taken as a case study, Webster's play shows that such tendencies were not just defiant remainders of a fading worldview. The work models a literary engagement with the sciences from the inception of long-eighteenth-century thought, and future scholarship might consider how other writers were shaping the technological and epistemological shifts of their time along with natural philosophers in an even more actively collaborative sense.

NOTES

1. Deborah Harkness, "'Strange' Ideas and 'English' Knowledge: Natural Science Exchange in Elizabethan London," in *Merchants & Marvels: Commerce, Science, and Art in Early Modern Europe*, ed. Pamela H. Smith and Paula Findlen (New York: Routledge, 2002), 138–141; Adam Max Cohen, *Shakespeare and Technology: Dramatizing Early Modern Technological Revolutions* (New York: Palgrave Macmillan, 2006), 4–5, 31.

2. Jonathan Sawday, *Engines of the Imagination: Renaissance Culture and the Rise of the Machine* (London: Routledge, 2007), 5.

3. Cohen, *Shakespeare and Technology*, 31.

4. Harkness, "'Strange' Ideas and 'English' Knowledge," 138; Cohen, *Shakespeare and Technology*, 19–31.

5. Michel Foucault, *The Order of the Things: An Archeology of the Human Sciences*, trans. anonymous (New York: Vintage Books, 1970), 20–25, 74.

6. Steven Shapin and Simon Schaffer's *Leviathan and the Air-Pump: Hobbes, Boyle, and the Experimental Life* (Princeton, NJ: Princeton University Press, 1985) offers perhaps the most famous contribution to the history of intersections between the epistemic dilemmas of natural philosophy and early modern technology, as well as Shapin's own *The Scientific Revolution* (Chicago: University of Chicago Press, 1996).

7. Ofer Gal and Raz Chen-Morris, *Baroque Science* (Chicago: University of Chicago Press, 2013), 7–8.

8. Wendy Beth Hyman, "'Deductions from Metaphors': Figurative Truth, Poetical Language, and Early Modern Science," in *The Palgrave Handbook of Early Modern Literature and Science*, ed. Howard Marchitello and Evelyn Tribble (London: Palgrave Macmillan, 2017), 31–32. Even so, those who argued most strongly for plain speech often used figurative language when it suited their purposes. Francis Bacon, Thomas Sprat, Thomas Hobbes, and John Locke all employ such speech in their treatises on natural philosophy, descriptions of the Royal Society, and explanations of experimental devices alongside their professed distaste for such ornament. See, for example, their arguments against rhetorical flourish: Francis Bacon, *The*

New Organon, trans. and ed. Lisa Jardine and Michael Silverthorne (Cambridge: Cambridge University Press, 2000), 41–42; Thomas Sprat, "History of the Royal Society," in *English Science, Bacon to Newton*, ed. Brian Vickers (Cambridge: Cambridge University Press, 1987), 162; Thomas Hobbes, *Leviathan*, ed. Richard Tuck (Cambridge: Cambridge University Press, 1996), 36; John Locke, *An Essay Concerning Human Understanding*, ed. Peter H. Nidditch (Oxford: Clarendon Press, 1975), 508. Scholars therefore debate whether their complaints are directed at figurative language itself or the heavy use of Latinisms by contemporaries, as explained in Ann Blair, "Natural Philosophy and the 'New Science,'" in *The Cambridge History of Literary Criticism*, ed. Glyn P. Norton (Cambridge: Cambridge University Press, 1999), 453.

9. Gal and Chen-Morris, *Baroque Science*, 11.

10. Ibid., 2–6.

11. Alina Alexandra Payne, introduction to *Vision and Its Instruments: Art, Science, and Technology in Early Modern Europe*, ed. Alina Alexandra Payne (University Park: Pennsylvania State University Press, 2015), 5; Susan Dackerman, introduction to *Prints and the Pursuit of Knowledge in Early Modern Europe*, ed. Susan Dackerman (Cambridge, MA: Harvard Art Museums, 2018), 19–20.

12. Payne, introduction to *Vision and Its Instruments*, 3.

13. Cohen, *Shakespeare and Technology*, 13–14, 16–17; Adam Max Cohen, *Technology and the Early Modern Self* (New York: Palgrave Macmillan, 2009), 13.

14. Cohen, *Shakespeare and Technology*, 4; Elizabeth Spiller, *Science, Reading, and Renaissance Literature: The Art of Making Knowledge, 1580–1670* (Cambridge: Cambridge University Press, 2004), 10–12, 14–16. The disciplinary boundaries between the arts and sciences that are so common in today's world did not exist in the same way for John Webster's contemporaries. As Cohen summarizes, scholars "have convincingly shown that the proto-scientific theory and practice often referred to in the period as *natural philosophy* was closely intertwined with humanistic study. While a significant gap may exist today between the humanities and the sciences, no such disciplinary gap existed during the early modern period" (*Shakespeare and Technology*, 4).

15. For recent scholarship that explores the possibility of viewing early modern English literary works in a more active relation with the scientific problems of their day, see Jenny C. Mann, *The Trials of Orpheus: Poetry, Science, and the Early Modern Sublime* (Princeton, NJ: Princeton University Press, 2021); Jenny C. Mann and Debapriya Sarkar, eds., "Imagining Early Modern Scientific Forms," special issue, *Philological Quarterly* 98, no. 1–2 (Winter/Spring 2019); Howard Marchitello and Evelyn Tribble, eds., *The Palgrave Handbook of Early Modern Literature and Science* (London: Palgrave Macmillan, 2017); and Carla Mazzio, "The Invisible Element in Art: Dürer, Shakespeare, Donne," in Payne, *Vision and Its Instruments*, 138–166.

16. For work on the relationship between anatomy theaters and drama, see Christian M. Billing, *Masculinity, Corporality and the English Stage, 1580–1635* (Farnham, UK: Ashgate, 2008); Hillary M. Nunn, *Staging Anatomies: Dissection and Spectacle in Early Stuart Tragedy* (Burlington, VT: Ashgate, 2005); and Penelope Meyers Usher, "'I Do Understand Your Inside': The Animal beneath the Skin in Webster's *Duchess of Malfi*," *Medieval and Renaissance Drama in England* 30 (2017): 105–125. Other scientific attention tends to focus on the occult arts and cases of lycanthropy, such as Lynn Maxwell, "Wax Magic and *The Duchess of Malfi*," *Journal for Early Modern Cultural Studies* 14, no. 3 (2014): 31–54; Mary Floyd-Wilson, *Occult Knowledge, Science, and Gender on the Shakespearean Stage* (Cambridge: Cambridge University Press, 2013); and Ellen Tullo, "Duke Ferdinand: Patient or Possessed? The Reflection of Contemporary Medical Discourse in John Webster's *The Duchess of Malfi*," *Medical Humanities* 36, no. 1 (2010): 19–22.

17. For such interpretations, see Michelle M. Dowd, "Delinquent Pedigrees: Revision, Lineage, and Spatial Rhetoric in *The Duchess of Malfi*," *English Literary Renaissance* 39, no. 3 (2009): 499–526; Ariane M. Balizet, "'Drowned in Blood': Honor, Bloodline, and Domestic Ideology in *The Duchess of Malfi* and *El Médico De Su Honra*," *Comparative Literature Studies* 49, no. 1 (2012): 23–49; and Andrea Henderson, "Death on the Stage, Death of the Stage: The Antitheatricality of *The Duchess of Malfi*," *Theatre Journal* 42, no. 2 (1990): 194–207.

18. *OED Online*, s.vv. "instrument," "experience," "experiment" (Oxford: Oxford University Press), https://www.oed.com.

19. John Webster, *The Duchess of Malfi*, ed. Michael Neill (New York: Norton, 2015), 15, 22 37.

20. Ibid., 42.

21. Ibid., 81.

22. While scholars do not typically view English literature in terms of the baroque, Webster's play has been studied through its aesthetics. For established connections between Webster and the baroque, see Ralph Berry, *The Art of John Webster* (Oxford: Clarendon Press, 1972). For relevant themes such as obsessions with refracted views, appearances that do match substances, and an unrelenting sense of mutability in the world, as well as their connections to specific technologies, see José Antonio Maravall, *La cultural del barroco: Análisis de una estructura histórica* (Barcelona: Editorial Planeta, 1975), 396; Christine Buci-Glucksmann, *The Madness of Vision: On Baroque Aesthetics*, trans. Dorothy Z. Baker (Athens: Ohio University Press, 2013), 8–10; Gilles Deleuze, *The Fold: Leibniz and the Baroque*, trans. Tom Conley (Minneapolis: University of Minnesota Press, 1993), 16–19; and Enrique García Santo-Tomás, *La musa refractada: Literatura y óptica en la España del Barroco* (Madrid: Iberoamericana, 2014), 44.

23. Édouard Glissant, *The Poetics of Relation*, trans. Betsy Wing (Ann Arbor: University of Michigan Press, 1997), 79.

24. Gal and Chen-Morris, *Baroque Science*, 11.

25. Webster, *The Duchess of Malfi*, 16.

26. Ibid., 17.

27. Ibid., 18.

28. Ibid., 54.

29. Ibid., 65.

30. Ibid., 56.

31. Ibid., 61, 31–32.

32. Ibid., 74.

33. R. B. Graves, "'The Duchess of Malfi' at the Globe and Blackfriars," *Renaissance Drama*, n.s., no. 9 (1978): 196–197.

34. Webster, *The Duchess of Malfi*, 76.

35. Margaret E. Owens, "John Webster, Tussaud Laureate: The Waxworks in 'The Duchess of Malfi,'" *ELH* 79, no. 4 (2012): 861–862.

36. Webster, *The Duchess of Malfi*, 77.

37. Lynn Maxwell, *Wax Impressions, Figures, and Forms in Early Modern Literature: Wax Works* (New York: Springer, 2019), 14–16, 134.

38. Ibid., 2.

39. Lucia Dacome, "Women, Wax and Anatomy in the 'Century of Things,'" *Renaissance Studies* 21, no. 4 (2007): 526, 528.

40. Lucia Dacome, *Malleable Anatomies: Models, Makers, and Material Culture in Eighteenth-Century Italy* (Oxford: Oxford University Press, 2017), 3; Maxwell, *Wax Impression, Figures, and Forms*, 10–13.

41. Maxwell, *Wax Impression, Figures, and Forms*, 17–23.

42. Webster, *The Duchess of Malfi*, 79.

43. Maxwell, *Wax Impression, Figures, and Forms*, 142.

44. Webster, *The Duchess of Malfi*, 79.

45. Quoted in Gal and Chen-Morris, *Baroque Science*, 44.

46. Jacques Lezra, *Unspeakable Subjects: The Genealogy of the Event in Early Modern Europe* (Stanford, CA: Stanford University Press, 1997), 103–105.

47. Quoted in Gal and Chen-Morris, *Baroque Science*, 43.

48. Ibid., 44.

49. Locke, *An Essay Concerning Human Understanding*, 363–364.

50. René Descartes, *Meditations on First Philosophy with Selections from the Objections and Replies*, trans. Michael Moriarty (Oxford: Oxford University Press, 2008), 22–23.

51. Locke, *An Essay Concerning Human Understanding*, 140–141, 117.

52. Vili Lähteenmäki, "Locke and Active Perception," in *Active Perception in the History of Philosophy: From Plato to Modern Philosophy*, ed. José Filipe Silva and Mikko Yrjönsuuri (New York: Springer, 2014), 234.

53. Jeffrey D. Burson and Anton M. Matytsin, introduction to *The Skeptical Enlightenment: Doubt and Certainty in the Age of Reason*, ed. Jeffrey D. Burson and Anton M. Matytsin (Liverpool: Liverpool University Press, 2019), 2.

54. Rainer Nägele, *Theater, Theory, Speculation: Walter Benjamin and the Scenes of Modernity* (Baltimore: Johns Hopkins University Press, 1991), 9.

55. Ibid., 6, 13.

56. Maxwell, *Wax Impressions, Figures, and Forms*, 145.

57. Webster, *The Duchess of Malfi*, 47.

58. Ibid., 106–107.

59. Susan Anderson, *Echo and Meaning on Early Modern English Stages* (New York: Palgrave Macmillan, 2017), 101. While Anderson primarily writes on the (dis)connections between Neoplatonism and sound, her description of the graveyard scene further underlines its interpretive instability for my analysis: "The play's frequent invocation of meaningful and nonsensical types of sound demonstrates the difficulty of distinguishing between them, establishing the possibility that the kinds of meaning we find in what we hear may, in fact, be our own inventions and misapprehensions. A listener's dilemma, then, is to determine which sounds to pay attention to, and how to interpret them, often without sufficient contextual clues to make a confident judgement" (ibid.).

60. Walter Benjamin, *The Origin of German Tragic Drama*, trans. John Osborne (Brooklyn, NY: Radical Thinkers, 2009), 209–210.

61. Webster, *The Duchess of Malfi*, 107.

62. Ibid., 115.

CHAPTER 2

Telling Time in the Fiction of Mary Hearne and Daniel Defoe

Erik L. Johnson

In late seventeenth- and early eighteenth-century Europe, and especially in England after 1680, rapid advances in timekeeping technology leant support to contemporary philosophies that treated time as a form of material extension analogous to extension in space. John Locke made the comparison explicitly in *An Essay Concerning Human Understanding*, introducing duration as "another sort of Distance, or Length, the *Idea* whereof we get not from the permanent parts of Space, but from the fleeting and perpetually perishing parts of Succession." In Locke's account, our abstract ideas of time and eternity, to say nothing of such measurements of duration as hours, days, and years, derive from reflection on the succession of perceptions in our own minds. By taking the direct experience of distinct, successive perceptions as the root origin of all our ideas about time, Locke accomplishes two things crucial to his larger epistemological project: (1) he equates time to physical motion and to change, treating it as essentially progressive in nature; and (2) he deprivileges the idea of eternity, treating it as a mode of time and a logical corollary of progressive extension (generated by adding "lengths of Duration to one another . . . *in infinitum*") instead of as a concept or experience of time that is somehow essentially different.[1] Like the isochronous "*Tick, Tick, Tick*" that had, Stuart Sherman writes, become a familiar means of describing the sound of a clock or watch in English by 1680, the Lockean view of time as based in succession is, if not exactly quantitative, fundamentally durational.[2] One can extend a count of minutes or a chain of ideas backward or forward—but in whichever direction one looks, new technologies and the empirical philosophy developing contemporaneously offer no escape from the present.

The attention that early eighteenth-century British authors of imaginative literature paid to how their characters experience and record the passage of time both reflects this intellectual and technological milieu and registers a great deal of uncertainty about it. Locke suggests, continuing his analogy of duration to spatial expansion, that by mapping our individual experiences in time and in place we can establish "known settled points" without which "all things would lie jumbled in an incurable confusion."[3] Yet in early eighteenth-century narrative, "insistence on the time process," a trait that Ian Watt sees as characteristic of a new, emerging novel form, does not always produce such reassuring results.[4] A watch may stop, changing a life forever; a makeshift calendar may omit entire days, throwing all the dates based on it into question. These events occur in fictions published in 1718 and 1719, pivotal early years in Watt's account in *The Rise of the Novel*, by the little-studied (and possibly fictitious) epistolary novelist Mary Hearne and by her contemporary Daniel Defoe. The detail of the temporal accountings that these two authors put on display does not so much establish order as reveal its fragility. Read against self-consciously modern early eighteenth-century dramas, like William Congreve's *The Way of the World*, that embrace a Lockean understanding of time as subjectively experienced but retrospectively universalizable, the fictions of Hearne and of Defoe show a deep suspicion of the modern, technologically mediated experience of time. In fact, considering Hearne's and Defoe's fictions in tandem suggests that many early eighteenth-century experiments in fictional narration pose philosophical challenges to Locke, defending an older, Augustinian understanding of time that centered on a felt dichotomy between the present and the eternal.

When and Where: Extension and the Materialization of Time

The technological culture of the early eighteenth century heavily favored analogies of time and space in Britain, since much of the energy of its ongoing horological revolution stemmed from the drive to develop a clock so reliable that it would be able to tell you not only when but where you were—by keeping time invariably enough to calculate the longitude at sea. By the seventeenth century, sailors could determine their latitude by observing the vertical position of astronomical reference points (such as the North Star in the Northern Hemisphere or the sun in the Southern Hemisphere). The earth's rotation makes calculating longitude a great deal more complicated, since objects in the sky appear to move horizontally even when one remains still. To adjust for this, one has to know how the local time differs from the time at the prime meridian—or one can try to estimate position based on the ship's speed and course.[5] Errors could have costs in lives and in treasure, and the steady toll of such disasters eventually led to government intervention.[6] Noting that theoretically promising means of measuring longitude remained imperfect "partly for want of money for trials and experiments," Parliament created through the Longitude Act of 1714 a Board of Longitude empowered to award the inventor or perfector of any such device or method with substantial sums—up to £20,000 for determination of longitude accurate to a half degree or thirty miles.[7] Though the clockmaker John Harrison would not develop a marine chronometer that met the Board of Longitude's standards until the 1760s, Isaac Newton's testimony to Parliament in support of the act made clear that "a Watch to keep Time exactly," one unaffected "by the Motion of a Ship, the Variation of

Heat and Cold, Wet and Dry, and the difference of Gravity," had moved from the realm of fantasy to that of theoretical technological possibility.[8]

The Longitude Act both emerged from and encouraged an understanding of time as a pure experience of technological mediation, a universalizable and homogeneous mathematical form not shaped by particularities of person or place. The horological revolution that had begun in Europe in the late seventeenth century focused on changes to two components of early modern mechanical clocks: the regulator, which kept time by dividing it into intervals, and the escapement, the mechanism through which the regulator moved the hands to indicate the time. In 1656, the Dutch mathematician Christian Huygens had raised the bar for accuracy in standing clocks by replacing the friction-prone foliot (rod) and balance wheel regulators of existing designs with pendulums. Though Galileo had previously suggested this idea, Huygens made it simple enough in the implementation that clockmakers both enthusiastically adopted the design for new clocks and retrofitted existing domestic timepieces en masse. In 1675, though there are sharper arguments about priority in this case, Huygens unveiled a parallel regulatory improvement for watches, the hairspring or spiral balance spring, which could release force from a weight a little bit at a time in regular intervals with far less variability than watch-sized versions of the pendulum.[9] At a stroke, or so the standard story goes, the best portable timepieces went from routinely gaining or losing fifteen to thirty minutes in the course of a day to general accuracy within five minutes under ordinary conditions on land. The innovations that catapulted England to "unchallenged preeminence" in clockmaking and watchmaking by 1680 were, however, more than refinements of Huygens's designs.[10]

If Huygens broke time down into finer grains—creating a chronometry in which, as Sherman argues, taking evidence from such inherently time-bound "diurnal forms" as diaries and periodical essays, the minute displaced the hour as the principal unit of time in much lived experience—British craftsmen made improvements that concealed and therefore naturalized the mechanism of clocks, weakening any links between clock time and natural diurnal rhythm.[11] Focusing much attention on the escapement, English clockmakers produced two new designs that greatly reduced internal wear and tear: the anchor escapement in the late seventeenth century (with Joseph Knibb, Robert Hooke, and William Clement as competing claimants) and the deadbeat escapement (by George Graham) in the early eighteenth. The Worshipful Company of Clockmakers, concerned about petitions for exclusivity of an early longitude projector, issued multiple broadsides in the 1710s popularizing the accomplishments of their member craftsmen, who, they boasted, "have Improved Watches . . . from being wound up every Twelve, to go Twenty four Hours, and now to a Week" and clocks "to go with a Pendulum Four Hundred Days, without once winding up."[12] For watches, the British developed the repeater, a feature that sounded out the current time, usually to the quarter hour, at the press of a button, which greatly enhanced the usefulness of portable timepieces at night. The repeater further abstracted clock time by removing it from any necessary association with the face of the clock.[13]

The new chapter "Of Identity and Diversity" that Locke added to the second edition of the *Essay* in 1694 outlined a vision of personal identity that further extends his treatment of time as pure duration. Locke's principal concern in this section is probative—he aims to establish a logical foundation on which persons can be held accountable for their actions. It is to that end that he insists that personal identity consists in "identity of consciousness," for "to punish *Socrates* waking, for what sleeping *Socrates* thought, and waking *Socrates* was never conscious of, would be no more of Right, than to punish one Twin for what his Brother-Twin did, whereof he knew nothing."[14] On the one hand, Locke engages in bold thought experiments in this section that seem to allow for highly psychologized experiences of time, asking, for instance, whether an individual can have different sleeping and waking personalities that constitute different persons and imagining that a consciousness might be transferred from one body to another. The Lockean notion of personal identity, however, is not so much psychological or even individual as it is durational. This identity consists, in the first place, of a succession of perceptions and, in the second, of the consciousness that arises from reflection on them. Duration provides both the impetus for that consciousness and its content. This definition of personal identity does not allow for any essential difference between subjects that would cause them to experience identical streams of perceptions differently.

Given contemporary neoclassical interest in using tightened unities of time to create heightened unities of action, it is not too surprising that the British stage at the turn of the eighteenth century would provide case studies illustrating that Huygensian chronometry and the Lockean conception of personal identity fit together, as both take temporal succession to be the fundamental organizing principle of experience. Perhaps inspired by the contemporary technological landscape, plays by Peter Motteux, John Dennis, William Congreve, and others following the Glorious Revolution tightened traditional versions of the unity of time by advertising that the time of their plotted action matched the time of performance perfectly, sometimes even including clocks or references to clock time to highlight the modernity of such simultaneity effects.[15] Congreve's *The Way of the World* demonstrates especially well the broad mapping functions that precision timekeeping can serve. The play gradually reveals varied and apparently conflicting commitments to be part of one harmonious design, once set in their proper chronological order. From its first scene, Congreve's comedy identifies the passage of time as a crucial thematic concern. In it, Mirabell asks a server for the time, receives the response, "Turn'd of the last Canonical Hour, sir," then (according to the stage direction) "*Looking on his Watch*," finds a more specific answer, exclaiming, "Ha? Almost one a Clock!"[16] The juxtaposition signals that Congreve's tightly plotted play will unfold according to two parallel but harmonized views of time. One model is traditional, bound to ritual and occasion, and intrinsic to the action; it is evoked here by the canonical hour, a pertinent reference because it signals that marriage schemes are afoot. The second, recorded by Mirabell's watch, is objective, mathematically precise, and allegedly impersonal—an indication that experience is mediated and, if not naturally well ordered, at least so orderable. These correspond roughly to plot and performance time, and Congreve keeps them closely synchronized, observing, in the words of Henry Home, Lord Kames, "a stricter unity of time during the whole play, than is necessary."[17] Keeping the whole to the space of a few hours in the afternoon, Congreve allows for plausible intervals between characters' exits and re-entrances, suitable for the business they state they will or have performed. More broadly, he illustrates the reach of durational thinking in a clockwork age.

The denouement of *The Way of the World* hinges on a new version of an old trope, the revelation of a secret identity, which we might in this case describe as delineated rather than revealed. Mirabell ultimately secures his marriage to the heiress Millamant and thwarts the villain Fainall's blackmail scheme by disclosing a financial agreement he made with Fainall's wife. Mrs. Fainall, whom we already know to be Mirabell's confidante, secretly put her estate in trust to Mirabell after the death of her first husband. The scene in which Mirabell produces the deed of trust explaining this recalls familiar tropes of recognition that hinge on the discovery of a previously unknown blood relationship between characters. Mirabell asks Fainall to read the inscription on the deed aloud—"*A Deed of Conveyance of the whole Estate real of* Arabella Languish, *Widow, in Trust* to Edward Mirabell."[18] We do learn a new name—her first one—and are able to identify the play's secondary heroine, with uncommon specificity for an eighteenth-century stage character, as Arabella Wishfort Languish Fainall. We already knew about each item in this jumble of identities. Arabella has more visible ties by blood and marriage than any other character onstage: she is Lady Wishfort's daughter, widow of a man named Languish, wife of the villain, and the hero's former lover. But now we can order these allegiances with precision. Drawn before Mrs. Fainall became Mrs. Fainall, the deed of conveyance establishes that her confederacy with Mirabell predates her marriage to Fainall and thus invalidates Fainall's claims to her estate. Against the background of the still-recent Glorious Revolution, *The Way of the World* resonates with an emerging Whig historiography in which forward progress becomes possible if we accept a view of human roles and commitments as durational rather than essential in nature.

Mary Hearne and the Limits of Temporal Accounting

In the first two decades of the eighteenth century, the increasing availability of clocks and watches made them prominent features of both Britain's material culture and its affective landscape. The French traveler Henri Misson had written in 1698 that London was a city characterized not so much by "horloges" (clocks) as by "horlogerie" (clockmaking), an art "en si grande vogue, que tout le monde a des montres, & qu'il y a de grandes pendules dans la plûpart des familles" (in such great vogue that everybody has watches, and that there are pendulum clocks in most households).[19] Looking beyond the capital, the historians Paul Glennie and Nigel Thrift find evidence of steadily expanding ownership of timepieces at all levels of wealth during the first four decades of the eighteenth century in probate inventories from Bristol, noting that by about 1720, "even the poorest group of inventories include some men with clocks or watches."[20] From the late seventeenth century, clock-based measurements appear with increasing frequency in surviving domestic documents such as recipes; in both published and manuscript recipe collections from 1675 to 1700, Glennie and Thrift report that instructions with specified intervals based on quarter, half, and whole hours (and sometimes intervals as precise as "half a quarter of an hour") steadily eclipse the early modern practice of using "task-based" instructions, such as saying a set number of Ave Marias, in order to time recipe steps.[21] New models of repeating watches continued to appear; most sounded hours and quarters, but by 1710, half-quarter and five-minute repeaters were also available. A surviving George Graham quarter repeater dating from 1719–1720 boasts a special feature, a "pulse piece" that, when pressed, causes the watch to vibrate, rather than ring out, the hour and quarter, allowing its user to check time discreetly.[22]

With the spread of chronometric consciousness into the domestic sphere, we should not be surprised to find that a timepiece and complicated affective responses to Huygensian timekeeping figure prominently in at least one amatory fiction, the first of two remarkable interconnected epistolary novels published in 1718 and attributed to a Mary Hearne, about whom little is known. Hearne, referred to as "Mrs. Hearne" in publishers' catalogs, "M.H." in the signed dedications to her fictions, and "Mary Hearne" in Edmund Curll's *A Compleat Key to the Dunciad*, was either a woman writer or Curll's conception of what a woman writer should be. Her first novel, *The Lover's Week; or, The Six Days Adventures of Philander and Amaryllis*, is dedicated to Delarivier Manley; its male lead shares his name with the hero of Aphra Behn's *Love-Letters between a Nobleman and His Sister* and at one point has "the famous Mrs. *Behn*'s Novels" on his table.[23] *The Lover's Week* and its direct sequel, *The Female Deserters*, are, in a sense, inherently critical of modern chronometric consciousness in that both follow characters' movement from highly scheduled urban life to pastoral retreat. Amit Yahav has recently argued that elements of romance (certainly a label that could fit Hearne's fiction) provide forms of resistance to and negotiation with Huygensian chronometry that eighteenth-century sentimental fiction ultimately reconstitutes into a new "sensibility chronotope."[24] *The Lover's Week*, however, provides more than a momentary escape from chronometric consciousness. By documenting a temporally regimented existence and then dramatizing its breakdown, Hearne's first novel raises political and philosophical questions about the entire enterprise of accounting for an individual life as a series of discrete successive moments.

As an epistolary fiction that consists of one long letter subdivided into six journal-like "days," *The Lover's Week* bears more than a few superficial similarities to Sherman's diurnal form; its narrator, Amaryllis, at first seems to provide "a continuous self-construction, a running report on identities both shifting and fixed," within the bounds of a "serial and closely calibrated temporality."[25] She writes to a confidante to explain her absence from London, and the status of the daily letters—whether she writes them at once retrospectively or incorporates prewritten material—remains unclear. But they generally follow an orderly diaristic form as though written a day at a time; they begin with Amaryllis rising and taking tea or breakfast around nine or ten, relate a series of engagements, and conclude with her retiring to bed around midnight. She vouches for the schedule by letting us know that she carries a timepiece and consults it frequently, remarking at various points that she took leave of her aunt "between five and six," that her maid "had not been gone out of my Chamber ten minutes by my watch, before she return'd," and that Philander was "the only Man that ever had obtain'd a quarter of an Hour's Discourse with me in private" (100, 101, 104). As the last remark indicates, such a thorough accounting of Amaryllis's whereabouts and activities serves to safeguard her reputation and her economic prospects; she resides with a single aunt who holds out hope that Amaryllis will marry an aristocratic suitor. She is also, apparently, correct to be vigilant, because when Philander seduces her, he does it by interfering with her ability to keep track of time.

In the third and by far the longest of *The Lover's Week*'s daily chapters, authors and characters alike chafe against timekeeping technology. Philander plans to invite Amaryllis in the evening and detain her through a series of excuses until it is too late to return home; he lays the ground, simply enough, by winding his own timepiece back, the better to mislead her. Hearne is concerned enough about plausibility that she offers multiple explanations for why this lie works, and as a result, the slippages of time that her heroine experiences are narratively overdetermined. Amaryllis notes, "looking on my own Watch, I found it did not go, though by what Accident it stop'd, I can't tell"—an "Accident" that remains unexplained. She also writes, in an aside that may have sufficed alone, "Time you know passes away quick in Company one likes, which I must own to you, I did that of *Philander's*" (106). In the aftermath, the heroine comes dislodged not once but twice from the socially acceptable chronometric grid that has organized her days. With the time past one, she accedes to Philander's argument that staying the night would be more prudent than returning home at an hour apt to attract scandal. The next day, this scenario recurs. Philander slips into her bedchamber in the night, and, though whether he does more than lie aside her sleeping form remains unclear, Amaryllis's rest is disordered enough that she does not take her morning tea on the fourth day until two in the afternoon, when "at so late an Hour, it was impossible for me to think of returning home" (112). With noise of her disappearance having begun to spread, she throws herself into well-ordered preparations for an elopement with Philander, as that now seems the most practicable course of action.

Amaryllis's stopped watch neatly represents the paradox at the center of Hearne's fiction, which depends on its heroine's fine-grained consciousness of time as it passes but elides, at its climax, any clear moment of decision or consent. The paradox is all the more striking because this elision is not either transcendent or tragic. At the end of *The Lover's Week*, Amaryllis describes her new life, lodged as Philander's mistress at his country retreat, as a happy one, preferable to any she could have enjoyed with the suitor her aunt preferred, but it is not a life outside of or above time. In fact, following the temporally distended events of the third night and fourth morning, she both recalibrates her narrative chronometry, settling back into a pattern of morning-to-midnight activity, and renews her participation in the comfortable material culture that her timepiece represents. She spends the fifth day in engagements with her "Milliner, Mercer, and Mantua-Maker" before moving and is pleased to learn that her new abode is "a quarter of an Hour's driving" from a market town (118–119). Instead of staging a simple contrast between diurnal form and timeless pastoral, *The Lover's Week* reveals the limitations of considering personal identity as the product of temporal succession by providing a thorough, moment-by-moment account of a pivotal week in its heroine's life that serves only to indicate her lack of awareness and agency in the events that change her circumstances most materially.

By including a story that forms a dark parallel to Amaryllis's own in *The Female Deserters*, the direct sequel to *The Lover's Week*, Hearne shows even more pointedly that a chain of successive, discrete, and more or less isochronous moments cannot capture the essential nature of female experience, so much of which is, in the eighteenth century, both passive and epochal, divided into chapters by life events, such as marriage, over which an individual may have little control. *The Female Deserters* conducts another experiment in epistolary form as, in a second long letter to her confidante, Amaryllis first continues her own story and then transcribes, in a nested, Chinese-box structure, the first-person amatory narratives of two new neighbors, her fellows in a small society united by experience of amorous "Circumstances" that have displaced them from the urban social grid (132). To a modern reader, one of these narratives, Calista's, may seem to be only Amaryllis's with the euphemisms removed.[26] She now lives as the mistress to a suitor who abducted her, drugged her, and went to bed with her in an unconscious state, leaving her with few socially viable options. Calista apologizes for the "Melancholy" tone of her story, but neither Hearne nor Amaryllis offers moral reflections on it, instead closing with emphasis on the "Rural Delights" that the women now enjoy (182–183). The "sleeping Drops" that override Calista's volition, however, serve a similar function, in her narrative, to the combination of deliberately altered and fatefully stopped watches in *The Lover's Week* (178). Put in the most general terms, these potent objects indicate how often events or conditions that the individual does not experience consciously nevertheless become principal determinants of personal identity.

Daniel Defoe and the Analogical Present

When Locke describes the ideas of duration, time, and eternity as ones we could arrive at through "those Sources of all our Knowledge, *viz. Sensation* and *Reflection* . . . as clear and distinct as many other," he dismisses as mystification a tradition of Christian thought based on the binary of temporal and eternal. Locke takes specific aim at St. Augustine, quoting a remark from book 11 of the *Confessions* ("the more I set myself to think of [time], the less I understand it") in order to contradict it by showing that duration and its modes are not so "abstruse in their nature." According to Locke, eternity is a mode of time, and we conceptualize it through the process of extension: "*The Mind having got the Idea of any Length of Duration, can double, multiply, and enlarge it,* not only beyond its own, but beyond the existence of all corporeal Beings."[27] According to Augustine, the idea of eternity arrived at through extension is false: "a long time . . . consisting of many passing shifts" differs qualitatively from eternity, in which "nothing passes, but everything is in the present." Among the three parts of time recognized in human language (past, present, and future), Augustine affords a special status to the present. Though the present is always in the process of passing into nonbeing, he writes, it is only in the present and in eternity that "being" is possible.[28] Though eternity is remote from human experience, we can conceptualize it, however imperfectly, through analogy with the present.

Defoe's fiction shows great sensitivity to the power of such analogical thinking about time. In *A Journal of the Plague Year,* set in 1665 and published in 1722, Defoe seems careful only to name London sites and streets that existed in both 1665 and 1720; in *The Fortunate Mistress,* popularly known as *Roxana,* he interweaves dates that place the action in the reign of George I with cameo appearances by figures of the Restoration (the banker Robert Clayton, the Duke of Monmouth, and, more ambiguously, a royal figure who may be Charles II).[29] In both cases, he emphasizes parallelisms between the historical experience of his characters and of his readers, offering us a vision of a narrator's essentially different present to which we can compare and contrast our own. If the reader of *A Journal of the Plague Year* can dimly envision the streetscape described when a desperate escapee from quarantine runs from one still-extant tavern to another (the Angel Inn, the White Horse, and the Pyed Bull), H.F.'s reflections on mortality and human nature become all the more resonant.[30] On the other hand, when Defoe drops the line "Written in the Year 1683" at the conclusion of *Moll Flanders* or when, following the dreamlike masquerade in which the heroine of *Roxana* receives her sobriquet, she pauses her own narrative to report that it will resume after a three-year hiatus (a private retreat, possibly with the unidentified king, that she is not at liberty to discuss), he uses chronological distance as a marker of difference, indicating that his characters' subjectivities may be comparable to, but are not continuous with, our own.[31]

Just as Hearne, writing in and about the domestic milieu in 1718, had developed a narrative that both depended on and critiqued modern chronometric consciousness as manifested in the improvement and spread of clocks and watches, so Defoe, in his landmark 1719 fiction *The Life & Strange Surprizing Adventures of Robinson Crusoe, of York,* developed a narrative that both drew on and qualified the durational model of experience at the heart of Locke's secular empirical philosophy. Defoe's fiction does evince the deep interest that Watt attributed to it in creating "a sense of personal identity subsisting through duration and yet being changed by the flow of experience."[32] At the same time, Defoe's heroes have radically idiosyncratic subjectivities, and his use of fictional chronologies that are familiar in their outlines but at best weakly correspondent to the world outside the text are a common means of signaling that idiosyncrasy. Defoe, it seems, intuitively grasped one of the major lacunae in Locke's thought—namely, just how it is, if all knowledge derives from individual experiences of sensation and reflection, that individual knowledge becomes communal.[33] One answer to the problem is Christian or Platonic—collecting and comparing experiences may allow us to perceive pieces of pre-existing truths whose veracity is either apparent or becomes so through their congruence.[34] Defoe's fiction posits, as an enabling condition of this approach, an Augustinian understanding of the present, as an unquantifiable, ineffable experience of time through which the individual can apprehend the eternal, but only by means of analogy.

We might read *Robinson Crusoe* as, in part, a religiously informed revision of Lockean philosophy. If Defoe reimagines the Garden of Eden through the lens of Locke's *Second Treatise of Government* when he describes Crusoe's cultivation of his island, he comes similarly close to the language of *An Essay Concerning Human Understanding* when he describes Crusoe's experience of time.[35] The dated journal of his first year on the island that Crusoe reproduces early in the narrative constitutes a remarkable formal innovation on Defoe's part—"the first time," Sherman notes, "a faux diary forms part of a larger fiction."[36] Its entries are also not chronologically reliable. Their dates presumably derive from the calendar that Crusoe tells us he began to keep by cutting notches on a post "about Ten or Twelve Days" after the shipwreck; the imprecision of that interval is curiously at odds with the fact that the calendar begins counting from the day of the wreck, September 30, 1659. One of the journal entries raises further questions about Crusoe's own count of days, for it includes a "*Note*," probably a retrospective addition: "I soon neglected my keeping Sundays, for omitting my Mark for them on my Post."[37] The most obtrusive indication that Crusoe's calendar is out of joint, however, comes in a dramatic incident that seemingly recapitulates one of the few biblical allusions to appear in Locke's *Essay*.

Illness and spiritual conversion dominate the late-June and early-July entries of Crusoe's journal, when the complementary promptings of reason and conscience drive the fever-wracked Crusoe to peruse one of the Bibles that survived the wreck and—something that, he writes, "I never had done in all my Life"—to kneel and pray. With this act and a quaff of rum, he sinks into oblivion, sleeping into the next afternoon and, perhaps, he speculates, even longer: "Nay, to this Hour, I'm partly of the Opinion, that I slept all the next Day and Night, and 'till almost Three that Day after; for otherwise I knew not how I should lose a Day out of my Reckoning in the Days of the Week, as it appear'd some Years after I had done. . . . But certainly I lost a Day in my Accompt, and never knew which Way" (1:108–109). This aside is one of several remarks in Crusoe's journal that only make sense if read as retrospective editorial insertions, though they are not distinguished typographically from what Crusoe calls the "Copy" of his journal (1:79). It also resembles Locke's conjectural account of the original human experience of duration. As further support for the dependence of our measures of duration on conscious experience, Locke notes that, though we generally have little difficulty retrospectively figuring intervals lost to sleep into our daily accounts of time, "if *Adam* and *Eve* (when they were alone in the World) instead of their ordinary Nights Sleep, had passed the whole 24 hours in one continued Sleep, the Duration of that 24 hours had been irrecoverably lost to them, and been for ever left out of their Account of time."[38] Whether or not the echo is intentional, it demonstrates how Defoe's self-sufficient fictional chronologies might act as a qualification to the Lockean view of duration. Crusoe on his island repeats what Locke considers to be a presocial experience of time, hardly isochronous and subject to gaps and distortions, of the sort that figure in the *Essay* mainly by way of thought experiment.

Notwithstanding this apparent interpolation, Crusoe's journal immediately proceeds to give dates after his fever breaks without offering any indication that they have been adjusted for publication; "I had no Fit the next Day," Crusoe writes, "this was the 29th" (1:109). Defoe's aim is not integration or harmonization. Coupled with the dated entries that ensue, the deictic "to this Hour" in the aside about the missing day, though it does signal some continuity with the past, also lets us know that multiple presents are simultaneously at play. Instead of taking these presents as successive, or even as organizable in a successive arrangement, Defoe presents conflicting narrative forms—the journal composed close to the action and the memoir composed long after—as parallel, warning us when first presenting the interpolated journal that it will repeat many "Particulars over again" (1:79). The effect is that time, for Defoe, is ineffable, or at least immaterial; duration is not a stable material substratum, equivalent to space, on which events can be plotted and organized without contradiction. Rather, Defoe urges an Augustinian qualification of the Lockean understanding of personal identity as shaped through the passage of time, implying that, while two bodies may not be able to occupy the same space, the same moment can easily include two (or more) presents. The moment we begin to think about the present, Augustine warned, it had already slipped into the past—but our apprehension of this very fleetingness means that we accept the relativity and boundedness of our own present(s).

Conclusion

In different ways, the fictions of Hearne and Defoe resist and qualify prevalent technologically based understandings of time as an idea derived from and subordinate to the simple experience of duration. By demonstrating how mastery of clock time fails to produce agency for a female protagonist, Hearne suggests that essentially different subjects may have essentially different experiences of time. Defoe makes a more extreme version of this case, focusing on subjects whose experiences of time are variously idiosyncratic. In doing so, he urges us to reconsider Locke's durational idea of personal identity, again implying that essential differences between subjects pre-exist and, to some degree shape, the subject's experience of duration. In the face of a homogenizing secular understanding of time that projects the present backward into the past and forward into the future, the fictions of Hearne and Defoe recall Augustine's belief that our present, and all presents, must be highly contingent, as they are but slivers of the eternal. These fictions suggest how older, spiritual understandings of time, if preserved by the novel, may continue to support critical comparative thinking about the present.

NOTES

I am grateful to Kit Kincade, the Defoe Society, and the American Society for Eighteenth-Century Studies for a valuable chance to present about the fiction of Mary Hearne. Blair Hoxby generously provided temporary access to his office and personal library while this work was in progress.

1. John Locke, *An Essay Concerning Human Understanding*, ed. Peter H. Nidditch (Oxford: Clarendon Press, 1975), 181–183, 193. In this respect, Lewis White Beck contrasts Locke to Isaac Newton, who posited eternity to be a sort of substratum for measurable time, "the infinite mathematical time that was before the world and will last after the world is destroyed." Beck, "World Enough, and Time," in *Probability, Time, and Space in Eighteenth-Century Literature*, ed. Paula Backscheider (New York: AMS Press, 1979), 117.

2. Stuart Sherman, *Telling Time: Clocks, Diaries, and English Diurnal Form, 1660–1785* (Chicago: University of Chicago Press, 1996), 1–2, 5–6.

3. Locke, *An Essay Concerning Human Understanding*, 199.

4. Ian Watt, *The Rise of the Novel: Studies in Defoe, Richardson, and Fielding*, 2nd ed. (Berkeley: University of California Press, 2001), 22.

5. Without mechanical timepieces up to that task, sailors gauged horizontal distance using manual aids such as marine sandglasses in combination with their estimates of a vessel's speed and course, a procedure known in English as "dead reckoning." The earliest reference to "dead reckoning" quoted in the *Oxford English Dictionary*, a seventeenth-century treatise on the virtues of the magnetic compass, contrasts a "dead reckoning" to the "true" one that a compass made possible in the calculation of latitude. Mark Ridley, *A Short Treatise of Magneticall Bodies and Motions* (London: Nicholas Okes, 1613), 147. In a short penultimate chapter on longitude, Ridley admits that "although . . . *Magneticall* needles . . . do not shew the longitude, . . . there is much helpe brought thereby for [its] knowledge" (149).

6. The Scilly naval disaster of 1707, in which navigational error led to the loss of Admiral Clowdisley Shovel's flagship the *Association*, three other vessels, and almost two thousand British sailors, including Shovel himself, became an inciting incident for passage of the act. The Scilly wreck provides a dramatic opening for Dava Sobel's popular history *Longitude: The True Story of a Lone Genius Who Solved the Greatest Scientific Problem of His Time* (New York: Walker, 1995), 11–14, though in fact it is not clear that miscalculation of longitude caused the tragedy, as Derek Howse notes in his authoritative *Greenwich Time and the Discovery of the Longitude* (Oxford: Oxford University Press, 1980), 45–47.

7. Great Britain, *An Act for Providing a Publick Reward for Such Person or Persons as Shall Discover the Longitude at Sea* (London: John Baskett, assigns of Thomas Newcomb, and Henry Hills, 1714). National longitude prizes had been offered before, notably by Philip II of Spain in 1567 (and again by his son Philip III in 1598), but Britain's rapid rise to preeminence in clockmaking and watchmaking in the late seventeenth century fostered new hopes of success.

8. *House of Commons Journal*, June 11, 1714, quoted in Howse, *Greenwich Time*, 51.

9. David Landes, *Revolution in Time: Clocks and the Making of the Modern World*, rev. ed. (Cambridge, MA: Harvard University Press, 2000), 127–128, on the pendulum clock; 136–138, on the balance spring and on the competing claims of Huygens and Robert Hooke. Cultural and religious networks contributed to the rapid spread of new practices across Protestant countries, particularly benefiting England, the republic of Geneva, and the Netherlands. Ahasuerus Fromanteel, a Dutch émigré to London and the first clockmaker to advertise pendulum clocks in England, in 1658, had, for instance, managed to apprentice his son John Fromanteel to Huygens's clockmaking partner in the Netherlands, Salomon Coster, in 1657, a year before Huygens's book about the pendulum regulator, *Horologium oscillatorium*, had even appeared in print. Clare Vincent and Jan Hendrik Leopold, *European Clocks and Watches in the Metropolitan Museum of Art: Highlights of the Collection* (New York: Metro-

politan Museum of Art, 2015), 86, 38 (item 17). France, which at first profited from Huygens's residence in Paris from 1665 to 1681, lost significant ground after Louis XIV's revocation of the Edict of Nantes in 1685 led to mass emigration by Protestant craftsmen. Roland Racevskis, *Time and Ways of Knowing under Louis XIV: Molière, Sévigné, Lafayette* (Lewisburg, PA: Bucknell University Press, 2003), 40–41.

10. The assertion is quoted from Carlo M. Cipolla, *Clocks and Culture, 1300–1700* (New York: Walker and Company, 1967) 69.

11. Sherman, *Telling Time*, 88–91.

12. See Worshipful Company of Clockmakers, *Reasons Humbly Offered . . . against the Bill, Entituled, a Bill for Securing to Mr. John Hutchinson the Property of a Movement Invented by Him for the More Exact Measuring of Time* (London, 1712). See also Worshipful Company of Clockmakers, *The Clockmakers' Farther Reasons against Mr. Hutchinson's Bill and Printed Reasons* (London, 1712).

13. Throughout the late seventeenth and early eighteenth centuries, the backup sundials that were once a common feature of watchcases and against which watch time could be checked steadily disappeared. Racevskis, *Time and Ways of Knowing*, 39. By 1731, the situation had so far reversed that an anonymous pamphlet discussed how difficult the reliability of watches made it to determine true solar time. "Twenty-four Hours of a good Watch are always exactly of the same Length; but Twenty-four Hours of the Sun are not exactly so," the author warns. He produces a table advising exactly how many minutes and seconds to wind a well-running watch back or forward to achieve a measure of solar time that reflects the differing lengths of days—thirteen minutes and twenty-two seconds forward from sundial time as index, for instance, to make up for the shortness of the day on January 16. *An Explanation of the Nature of Equation of Time, and Use of the Equation Table for Adjusting Clocks and Watches to the Motion of the Sun* (London: F. Clay, 1731), 5, appendix.

14. Locke, *An Essay Concerning Human Understanding*, 342.

15. Examples include Congreve's plays, John Dennis's *A Plot, and No Plot* (1697), and Peter Motteux's *Beauty in Distress* (1698), *The Temple of Love* (1706), and *Love's Triumph* (1708). Showy observance of "perfect" unities of time became common enough that George Farquhar complained about them in *A Discourse upon Comedy*, in *The Works of George Farquhar*, ed. Shirley Strum Kenny, 2 vols. (Oxford: Clarendon Press, 1988), 2:270; and John Gay and Alexander Pope satirized them in *Three Hours after Marriage* (1717), in *John Gay: Dramatic Works*, edited by John Fuller (Oxford: Clarendon Press, 1983), 207–263.

16. William Congreve, *The Way of the World*, in *The Works of William Congreve*, ed. D. F. McKenzie, 3 vols. (Oxford: Oxford University Press, 2011), 2:107–108.

17. Henry Home, Lord Kames, *The Elements of Criticism*, 6th ed., ed. Peter Jones, 2 vols. (Indianapolis: Liberty Fund, 2005), 2:429.

18. Congreve, *The Way of the World*, 2:221.

19. Henri Misson, *Memoires et observations faites par un voyageur en Angleterre* (The Hague: Henri van Bulderen, 1698), 239; translations mine.

20. Paul Glennie and Nigel Thrift, *Shaping the Day: A History of Timekeeping in England and Wales, 1300–1800* (Oxford: Oxford University Press, 2009), 125. To the extent they can determine it, Glennie and Thrift assert that this pattern holds for England as a whole, with the diffusion of clocks and watches "relatively rapid compared with many consumer goods, especially given their relatively high cost" (163–164). They also note that the emphasis that most horological studies have placed on privately owned clocks and watches neglects the improvement of public clocks, in particular the addition of clocks to civic buildings and to poorer parishes. Existing historical literature, they contend, with its emphasis on the private, potentially idiosyncratic experience of time, "chronically underestimates the temporal competencies of eighteenth-century people and environments" (116).

21. Ibid., 223–224.

22. Vincent and Leopold, *European Clocks and Watches*, 156 (item 33).

23. Page citations for Hearne's novels, here and in subsequent in-text parentheses, are to the edited texts in Joan Rose Reteshka, "An Edition of Mrs. Hearne's 'The Lover's Week' and 'The Female Deserters'" (PhD diss., Duquesne University, 1998). These heavy-handed nods to

popular authors and the rather mechanical plug for her two novels in *A Compleat Key* have led some critics to suggest that Hearne was an invention of Curll's. The distinctiveness and unity of her two novels and the fact that there are only two lead me to believe that she existed, though "Mary Hearne" may of course be a pseudonym. *The Lover's Week*, published in 1718 by E. Curll and R. Francklin, appeared in a second stand-alone edition within six months of publication and in 1724 was serialized in *The Original London Post*. Robert Adams Day, in *Told in Letters: Epistolary Fiction before Richardson* (Ann Arbor: University of Michigan Press, 1966), labeled it "one of the most . . . significant pieces of English fiction produced before 1740" (177). Her second fiction, *The Female Deserters*, appeared in November 1718 under the imprint of J. Roberts and was published in a combined edition with *The Lover's Week* under the title *Honour the Victory and Love the Prize* in 1721. For more on publication history and authorship, see esp. Paul Baines and Pat Rogers, *Edmund Curll, Bookseller* (Oxford: Clarendon Press, 2007), 120, 342n29; Robert Ignatius Letellier, *The English Novel, 1700–1740: An Annotated Bibliography* (Westport, CT: Greenwood Press, 2002), 484; Kathryn R. King, "The Novel before Novels (with a Glance at Mary Hearne's Fables of Desertion)," in *Eighteenth-Century Genre and Culture: Serious Reflections on Occasional Forms, Essays in Honor of J. Paul Hunter*, ed. Dennis Todd and Cynthia Wall (Newark: University of Delaware Press, 2001), 36–57; and Reteshka, "An Edition," 4.

24. Amit S. Yahav, *Feeling Time: Duration, the Novel, and Eighteenth-Century Sensibility* (Philadelphia: University of Pennsylvania Press, 2018), 9, 14. Yahav argues that the sensibility chronotope renders duration qualitative through the incorporation of "off-the-clock" breaks into novelistic narrative.

25. Sherman, *Telling Time*, 8–9.

26. Hans Turley, in "The Anomalous Fiction of Mary Hearne," *Studies in the Novel* 30, no. 2 (1998): 146, for instance, argues for a moralistic reading of Hearne's fictions and insists that the second novel, as well as "the noise of all the other heroines with whom readers are familiar," effectively undermines the happy ending of the first. Reteshka, in "An Edition," discusses Hearne among other precursors to Samuel Richardson in an appendix on "the drugged rape scenario" (195–198).

27. Locke, *An Essay Concerning Human Understanding*, 181, 197. The parenthetical reference to Augustine is quoted from Locke.

28. Augustine, *Confessions*, trans. Carolyn J. B. Hammond, 2 vols., Loeb Classical Library 26–27 (Cambridge, MA: Harvard University Press, 2014–2016), 2:211–213, 2:217–219.

29. Paul K. Alkon, *Defoe and Fictional Time* (Athens: University of Georgia Press, 1979), 49–50, 52–54. Manuel Schonhorn identifies and dates ninety topographical references in "Defoe's 'Journal of the Plague Year': Topography and Intention," *Review of English Studies*, n.s., 19 (1968): 387–402, duly credited in Alkon, *Defoe and Fictional Time*, 49n33. Alkon frames *A Journal of the Plague Year* and *Roxana* as complementary rather than contrasting examples of Defoe's attitude toward fictional time and anachronism. By mentioning only buildings that existed in both 1665 and 1720, *A Journal of the Plague Year* avoids anachronism in a technical sense but in actual effect blurs distinctions between present and past. On dates and historical figures in *Roxana*, see esp. David Blewett, "The Double Time-Scheme of *Roxana*: Further Evidence," *Studies in Eighteenth-Century Culture* 13 (1984): 19–28; Judith Sloman, "The Time Scheme of Defoe's *Roxana*," *English Studies in Canada* 5, no. 4 (1979): 406–419; and Rodney M. Baine, "Roxana's Georgian Setting," *SEL* 15, no. 3 (1975): 461–464.

30. Daniel Defoe, *A Journal of the Plague Year*, ed. Paula R. Backscheider (New York: Norton, 1992), 61.

31. Daniel Defoe, *The Fortunate Mistress; or, a History of the Life and Vast Variety of Fortunes of Mademoiselle de Beleau . . . Being the Person Known by the Name of the Lady Roxana, in the Time of King Charles II*, 2 vols. (Oxford: Basil Blackwell / Shakespeare Head Press, 1927), 1:212.

32. Watt, *The Rise of the Novel*, 24.

33. Nancy Armstrong, in "The Other Side of Modern Individualism: Locke and Defoe," in *Individualism: The Cultural Logic of Modernity*, ed. Zubin Meer (Lanham, MD: Lexington

Books, 2011), 111–119, puts the problem in political terms, noting that "Enlightenment intellectuals found it notoriously difficult to explain how Locke's individual could join with others to form a cohesive community" (112), but it is no less an epistemic one. Amit Yahav, in "Time, Duration, and Defoe's Novels," *Partial Answers: Journal of Literature and the History of Ideas* 6, no. 1 (2008): 33–56, points out that Berkeley's treatment of ideas (including ideas of time) as mind-dependent and Leibniz's insistence that only a robust a priori framework of logical and mathematical concepts allows us to generalize from our observations and experiences are both important counterpoints to "the ease by which [Locke] moves from individual experience to objective measure" (37).

34. This is the journalistic project that Defoe lays out in the first issue of his periodical *A Review of the Affairs of France* (no. 1, February 19, 1704), in which he promises not so much reportage as "needful Rectifications" by comparing reports and, "both Sides being Examin'd, the Particulars referr'd to the general Opinion of all Men." Religious faith supports Defoe's belief that comparative reading will lead to discovery and that the quest for truth will have an endpoint. He describes his hoped-for results in strongly Miltonic language, pledging, "Nor shall we Embroil our selves with Parties, but pursue the Truth; find her out, when a Crowd of Lyes and Nonsence has almost smother'd her, and set her up, so as she may be both seen and heard." Daniel Defoe, *A Review of the Affairs of France*, ed. John McVeagh, 9 vols. (London: Pickering and Chatto, 2003), 1:7–8. Cf. Milton's *Eikonoklastes*: "It shall be ventur'd yet, and the truth not smother'd, but sent abroad, in the native confidence of her single self, . . . to finde out her own readers." In *Complete Prose Works of John Milton*, ed. Merritt Y. Hughes, vol. 3 (New Haven, CT: Yale University Press, 1962), 339.

35. Rebecca Bullard, "Politics, History, and the *Robinson Crusoe* Story," in *The Cambridge Companion to "Robinson Crusoe,"* ed. John Richetti (Cambridge: Cambridge University Press, 2018), 89–92, details parallels between passages in *Crusoe* and Locke's *Second Treatise.*

36. Sherman, *Telling Time*, 226.

37. Daniel Defoe, *The Life & Strange Surprizing Adventures of Robinson Crusoe, of York*, 3 vols. (Oxford: Basil Blackwell / Shakespeare Head Press, 1927), 1:72, 82. Subsequent parenthetical references in the text are to this edition.

38. Locke, *An Essay Concerning Human Understanding*, 183.

CHAPTER 3

The Technology and Theatricality of *Three Hours after Marriage*'s "Touch-Stone of Virginity"

Thomas A. Oldham

Within a few weeks of the 1717 premiere of John Gay, Alexander Pope, and John Arbuthnot's *Three Hours after Marriage*, a response appeared in print. Written by "E. Parker, Philomath," the *Complete KEY* features, among its many observations, this striking critique: "*Lubomirski's Liquor for the Tryal of Virginity.* Is an Incident stole out of an old Play, Printed in 1653. (somewhat applicable to these Scribblers) call'd, 'The *Changeling*, a Tragedy. Written by *Thomas Middleton*, assisted by Mr. *Rowley*.'"[1] Parker noticed that the Scriblerians had written a scene remarkably like one that had graced London's stage nearly a century earlier. Licensed for performance in 1622, *The Changeling* features Beatrice-Joanna, a woman given a concoction by her husband to test her virginity. Having studied how to pass the test (by yawning, sneezing, and laughing), Beatrice fools Alsemero into thinking she is chaste, which the audience knows is false. Hers is merely a simulacrum of virginity, ultimately exposed by the series of events leading to her tragic death. *Three Hours after Marriage*, however, takes this scene and renders it comic. Susanna Townley, drinking a similar potion, fails to faint, sneeze, or produce a large red spot of infamy on her cheek—all telltale signs of a deflowered woman. Again, dramatic irony informs the audience that she is no virgin. The pseudoscientific stunt was a sham all along: Susanna's lovers duped her husband, an ostensibly learned doctor, with their test. What was once tragic now provokes laughter.

While *The Changeling*'s easily falsified chastity may strike some as comic, the play's macabre violence is serious.[2] This virginity test should be taken seriously as well, as it proves accurate for the truly virginal Diaphanta and is "very much in touch with tradition but modified to be stageworthy."[3] The play's scientific and technological context was one in which alchemy had not fully surrendered to the scientific method, with Alsemero's work related to both the historical genre of the Renaissance closet[4] and the writings of Antonius Mizaldus, sixteenth-century astrologer and doctor.[5] The virginity test does not appear in Mizaldus's *De Arcanis Naturae* (the book mentioned in the play), but there are such tests in his *Centuriae IX. Memorabilium*, continuing a centuries-long tradition.[6] Supposed signs of virginity traditionally included blood at first intercourse, tightness/constriction of the vagina/cervix/uterus, distinct volume/odor/sound of urine, a propensity to urinate after consuming various compounds, the shape of the breasts, and various changes in complexion. Many texts caution that these signs can be falsified, and some even provide instructions on how to do precisely that.[7] Lest one think that such ideas would have been laughable by 1622, Johann Jacob Wecker perpetuated several techniques of urinalysis as late as 1660.[8] The period also saw a number of attempts to establish virginity via physical examination, although despite a proliferation of anatomical texts, understanding of the hymen was decidedly uncertain.[9] Of special interest to historians is the case of Frances Howard, whose divorce from the Earl of Essex centered on his impotence and her claims to virginity, allegedly proved by physical examination. This has direct textual ties to *The Changeling*.[10]

Between *The Changeling* and *Three Hours after Marriage* were ninety-five years of intellectual and technological advances that accompanied the height of the scientific revolution and the beginning of the Enlightenment, including new medical technologies and wider dissemination of anatomical knowledge. This recontextualization of the falsified, pseudoscientific virginity test from a Jacobean tragedy to a Scriblerian satire represents changing attitudes toward science and the female body. With its satirical treatment of husband and wife, *Three Hours after Marriage* simultaneously considers and undercuts advances in medical science, technology, and women's agency.

The seventeenth and eighteenth centuries have long held significance for historians of sex and gender. Thomas Laqueur, in his landmark *Making Sex*, pinpoints this era as key to the transition from the one-sex to the two-sex model.[11] While this interpretive timeline has been critiqued by Karen Harvey, it remains highly influential.[12] Sweeping assessments from Laquer and Michel Foucault have fundamentally altered the way historians understand the period, but perhaps more useful to understanding the individual body is the work of Mary E. Fissell, whose focus on English vernacular texts is "more particular, and in this case a smaller canvas is a messier canvas."[13] Fissell identifies small ways in which the Reformation, the Civil War, and the Restoration affected the narratives surrounding the female body. Many of these changes were textual: "In the midst of 'a world turned upside down,' some writers turned to the human body itself as a possible producer of stable truths."[14] The downside to this was the possibility that texts could "propose that women were inherently inferior to men, implying that sexual subjugation was grounded in natural fact."[15]

English vernacular texts approached sex and generation with the conventions of various categories: midwifery texts, natural histories, and books of wonders.[16] All of these genres addressed virginity in significant ways, but with no consensus. Combined with the apparent objectivity of detailed descriptions and technologically sophisticated engravings, this ambiguity renders virginity a site of contestation. Texts like Helkiah Crooke's *Mikrokosmographia* (1615), which Kassell calls the "first comprehensive English anatomy,"[17] is typical in its emphasis on the hymen: "the entrance, the piller, or locke, or flower of virginity. For being whole it is the onely sure note of unsteyned virginity."[18] This is also true in a practical text like Nicholas Culpeper's *A Directory for Midwives* (1651): "four Caruncles or Fleshy Knobs" that are "round in Virgins, but hang flagging when Virginity is lost . . . joyned together by a thin and sinewy skin, or *Membrana* . . . called the *Hymen*." This hymen provides "a certain note of Virginity where ever it is found, for the first act of Copulation breaks it."[19] Other books were of a decidedly less scientific nature. While the work of Giovanni Sinibaldi (appearing in English in 1658 as *Rare Verities: The Cabinet of Venus Unlocked and Her Secrets Laid Open*) defines loss of virginity as hymeneal rupture (while also cautioning it may be broken in many ways), it also features sections closer in tone and content to *The Changeling*.[20] Dubious instructions identify "any Signs of corrupted virginity," including impudence, a forked nose, an elongated neck, the inability to endure the smell of a lily, and attracting bees, while also providing tricks to cause incontinence or ever-faithful embraces.[21] Corinne Harol, in her extensive study of Enlightenment virginity, describes the overall trend as one of empirical epistemology subjecting "the hymen to the same objectification as other body parts, contributing to the demystification of virginity."[22]

Many of the examples from the period follow similar patterns. Both Thomas Chamberlayne's *The Compleat Midvvife's Practice Enlarged* (1659) and *The Midwives Book* (1671) by Jane Sharp (the rare midwife who published herself) are skeptical, providing alternate explanations for a broken hymen.[23] While Sharp is celebrated for inverting male-female rhetoric and normalizing the female body, the sheer number of male-authored texts minimizes such an effect.[24] Thomas Gibson's *The Anatomy of Humane Bodies Epitomized* (1682) vaguely describes a virgin's uterus as unique in "substance," "bigness," and "cavity."[25] He describes a virgin's vagina as "so strait," yet he cautions that the lack of the "signs of Virginity" are not definitive proof that "Virginity is also wanting."[26] The popular, pseudonymous *Aristotele's Master-Piece* (1684) describes the female anatomy as possessing "4 Caruncles, or little Buds like Mirtle-berries, which in Virgins are full and plump," which are "joyned or held together by little Membranes."[27] These membranes, "being once delacerated, denote devirgination."[28] Robert Barret's *A Companion to Midwives, Child-Bearing Women, and Nurses* (1699) claims that this hymeneal tradition is "not to be doubted, we have such great Authorities for it," even though the sign may not be "perfect."[29] William Cowper's *The Anatomy of Humane Bodies* (1698) cautions that the hymen appears differently at different ages and may be broken "in Coitu or otherwise."[30] Michael Etmullerus's *Etmullerus Abridg'd* (1703) describes falsified virginity, "obtain'ed by straitning the Genitals, and retrieving their natural Tone: which is effected by Baths and Fomentations prepar'd of Astringents" made of various roots.[31] Nicolas Venette's *The Mysteries of Conjugal Love Reveal'd* (1707) is remarkably forward-thinking: "there is no certain Sign of a Maidenhead, nor of the Ravishment of a Virgin," and such claims "are for the most part equivocal, and uncertain," largely based on the performance of certain attitudes and behaviors.[32] One page, later, however, he provides a recipe for how "a counterfeited Maidenhead may be discover'd."[33] John Marten's *Gonosologium Novum* (1709) reiterates that the hymen is not a reliable indicator of virginity and details a way to deceive husbands by using a fish-bladder of blood.[34]

By 1717, the year of *Three Hours after Marriage*, Culpeper's *Directory* had appeared in several editions. In the first printing, he cautioned that the hymen "may be broken without Copulation, as it may be gnawn asunder by defluxion of sharp humors."[35] The expanded 1716 version equivocates further: the hymen "is not found in all Virgins because some are very lustful, and when it itcheth, they put in their Finger or some other thing and break the Membrane: Sometimes the Midwives break it."[36] Not all virgins will bleed on first copulation (as both Moses and "African Custom" hold) for a variety of reasons, including "if the Man's Yard be small."[37] Despite this relative scientific accuracy, it also utilizes antiquated Galenic, humoral terms.[38] These texts share the ambivalent anxiety present in both *The Changeling* and the overall "Restoration crisis in paternity."[39] There is certainty and ambiguity; the sources of knowledge are simultaneously antiquated and up-to-date, direct and indirect. Even Pseudo-Aristotle, not a soberly scientific writer, attempts to sound definitive, detailing the story of a man tried for rape who was cleared by an examination by a "skillful Chirurgeon and two Midwives," who concluded that the supposed victim was still a virgin.[40]

The reference to surgeon and midwife highlights one aspect of the period: "Medical understanding of the body was informed by and enacted through medical practice. The best histories of early modern medicine treat ideas and practices as two sides of the same coin."[41] This had not always been the case; Michael Stolberg examines the "limited role of physical examination in traditional bedside medicine" for much of the premodern era.[42] Consequently, until roughly the eighteenth century, parturition was solely the realm of the midwife. Male surgeons typically only intervened when complications endangered the mother's life. Not only did midwives help women in labor; their knowledge of and authority over bodies allowed them to serve myriad functions, from medieval virginity tests to the Frances Howard case.[43] The eighteenth-century physician John Leake acknowledged midwifery as "the *Medicina forensis*, or that branch of the Science contributing to the public administration of justice, in what relates to *virginity, pregnancy*, and the *natural period of uterine gestation*."[44] Delivering babies and assessing virginity dovetailed as midwives were charged with ensuring inviolate lines of succession through proper births. Men could gain knowledge from books, but midwives exerted control properly with direct experience and agency in the matter. This helped create what Adrian Wilson terms "a collective culture of women" surrounding childbirth.[45]

Over time, gender relations changed as the midwife slowly ceded authority to her male counterpart. The reach of these new man-midwives, who "made reproductive knowledge available to all," extended far past the birthing chamber.[46]

Increasingly seen as more objective, rational, and fit, they lectured, published texts, testified as to whether criminals were pregnant (and thus exempt from execution), and, significantly, established paternity.[47] Caroline Bicks notes that the traditional midwife had a profound effect on patriarchal lineage and masculinity, largely because her "potential manipulations of virgins (or of fallen women wishing to appear virginal) intersect . . . with larger issues about the maintenance of clearly defined and recognizable categories by which female bodies might be contained and circulated in the marital marketplace."[48] Man-midwives now possessed this knowledge of women and offspring, ultimately shaping the resulting connections between sex and money. For Roy Porter and Lesley Hall, the economic implications of sex are clear, as it "was inscribed within rules and norms, and it was subject to an exchange economy in which the ultimate regulator was the balance between production and reproduction."[49] Assuring a husband that his wife was sleeping only with him was more knowable—and more important—than ever. For Tassie Gwilliam, "the presence of virginity in eighteenth-century marketplaces—of texts, marriage, or prostitution"—means that "counterfeit maidenheads devalue virginity in the same way that fake money devalues coinage."[50] The equivocations present in the anatomical literature, however, belie the false sense of security that scientific knowledge seemed to provide. When the commodification of female sexuality is taken into account, the ambiguous nature of virginity displayed in anatomical texts and midwifery manuals meant that entire economic systems based on patrilineal inheritance were at stake: "the virgin anatomy becomes a way of working out power/authority issues between men."[51] While concerns over cuckoldry and bastard children had always existed, the man-midwife and the (male) surgeon seemed to establish greater patriarchal control.

This shift was assisted, at least in part, by a new technology, cloaked in secrecy yet exhibited with theatricality:

> The one thing that Peter [Chamberlen] III had neither delivered nor unfolded was the Secret—the secret invention, probably of Peter the Elder, which enabled the Chamberlens to deliver more women more effectively than any obstetrician before them. The Chamberlens had always gone to fantastic lengths to keep their secret. They are said to have arrived at the house of whichever woman it might be in a special carriage. With them was a huge wooden box adorned with gilded carvings. It always took two of them to carry the box and everyone was led to believe that it contained some massive and highly complicated machine. The laboring woman was blindfolded lest she should see the Secret. Only Chamberlens were allowed in the locked lying-in room, from which the terrified relatives heard peculiar noises, ringing bells, and other sinister sounds as the Secret went to work.[52]

The Chamberlen "Secret" was the obstetric forceps. The development and popularization of the forceps, from roughly the mid-seventeenth to early nineteenth century, marked a change in the practice of childbirth and, consequently, in the relationship of men to women's bodies.[53] This anecdote illustrates the difficulty in understanding the invention's history: for at least a century, the nature of the forceps was a proprietary secret guarded for the sake of the Chamberlen family's fame and fortune. The key features of their forceps show the careful innovation incorporated into the design: two separate, fenestrated blades with slight curvature to grasp the head gently, but the timeline of these adaptations, and who developed them, is unclear.[54] What is clear is the immense benefit gained from the device, and this success led to other boastful theatrics from the Chamberlens, from College of Physicians politics to the doomed pregnancy of a rachitic French "dwarf."[55] By the time Edmund Chapman included a description of similar forceps in 1733's *Essay on the Improvement of Midwifery*, many people assumed that the heirless Hugh Chamberlen sold the family secret, though the prototypes were not found until 1813, hidden underneath the floorboards of Peter Chamberlen's house.[56]

However the Chamberlen secret spread, its impact was extensive. Leake's 1787 edition of his midwifery book touts its "DESCRIPTION of the Author's NEW FORCEPS, Illustrated with elegant COPPER PLATES" alongside a condescending view of midwifery's past: "Women in Labor were attended by their own sex, who gave their assistance without method, being rather directed by necessity than skill or choice; hence, the progress and improvement of this Art must have been extremely slow."[57] A new age dawned, however, that had profound effects on gender relations. Bettina Wahrig sees the introduction of these "tools of exclusion" as the moment when midwifery became grounded in "a rational and mechanical explanation of life": without access to the proper educational and financial resources, women were deemed unfit to handle those tools.[58] This forceps-driven tension between men and women in the birthing chamber was commonplace enough to be satirized in Laurence Sterne's *Tristram Shandy*.

By the early 1700s, men were attending at an increasing number of healthy births.[59] The rise of the man-midwife was the result of complicated historical forces, political influence, and class dynamics, all aided and abetted by the new technology of the forceps.[60] An equally powerful instrument for change in gender relations, though, might be performance itself, as the Chamberlens' theatricality suggests. Performance theory has influenced the study of sex and gender for decades;[61] thus, any difficulty of proving virginity goes beyond the inscrutability of the individual hymen and amplifies the performativity of all women. Scholars have used such language to describe *The Changeling*, saying its virginity is a "re-enactment, . . less a state of sexual innocence than a performing art,"[62] as well as being "deeply dependent upon the stories women told about their sexual status and on how they acted these stories out."[63] In other words, Beatrice-Joanna's and (as we shall see) Susanna Townley's virginity is performative. If virginity is not a physical, bodied state, these characters are able to create through words and repeated signs the very thing under question. In the case of *Three Hours after Marriage*, what could have been a tragedy of virginity becomes farcical; deadly serious, arcane alchemical knowledge gives way to a performative scientific parody. Yet this parody is not necessarily progressive or feminist. While the classic line about the phenomenology of the theatre ("in theater there is always a possibility that an act of sexual congress between two so-called signs will produce a real pregnancy")[64] takes on new resonance in this play, feminist theatre scholars argue against any neutrality of this phenomenological power: "Placing women in a representation always connotes an underlying ideology and presents a narrative driven by male desire that effectively denies women's subjectivity."[65] This parallels Fissell's analysis that Culpeper's *Directory for Midwives* "sexualizes the reading process, implicitly making women's bodies into objects to be consumed by men."[66] *Three Hours after Marriage* shows the theatre to be an instrument with precisely these effects.

Known primarily as the play in which rival suitors dress up as a mummy and a crocodile, *Three Hours after Marriage* shows the Scriblerians playing virginity anxiety for laughs, discharging their satire on a favorite hobbyhorse: the foolish pedant, Dr. Fossile, physician and antiquary, is a stand-in for Dr. John Woodward, physician and antiquary.[67] Al Coppola indicates, however, that "it is crucial to recognize that the Fossile of the farce is not really John Woodward." The Scriblerians, he argues, overlay aspects of a classic figure of satire, the virtuoso, on top of Woodward, creating a figure "who seems much more at home in the sixteenth or seventeenth centuries than in the eighteenth century."[68] Thus, it is logical that Coppola compares the play to a Jonsonian comedy of humors; it is also appropriate to remember its similarity to the Jacobean *The Changeling*, as Gay's aforementioned contemporary duly recognized.[69]

The plot is simple: Fossile marries a young bride but is disabused of any illusions that she is virginal when he learns that she is already married and the mother of a child. Supporting characters include two additional suitors, a mediocre female playwright, and a pompous literary critic. Before the ending revelations, Fossile administers the virginity test to Susanna at the height of his obsession with her chastity. Whereas *The Changeling*'s Beatrice-Joanna counterfeits the indicators of true virginity, Susanna does not need to dissemble, not because she is pure but because the test is pure hokum. Fossile is fooled into believing the scientific-sounding *Lapis Lydius Virginitatis* (Touchstone of Virginity) by a disguised rival. Fossile believes, but the audience is in on the joke. A century of scientific advancement rendered what had once been a tragic scene satirical. Beneath the humor, however, lies the truth of science-enabled male hegemony: Fossile's pseudoscientific credulity may be an object of ridicule, but Susanna reveals the debasement of female agency.

A knowing bawdiness is part of the play's appeal: Susanna furtively entertains her paramours Plotwell and Underplot via a series of farcical events, including a suggestive scene in which Plotwell hides under her petticoat. Calhoun Winton argues that the double entendres, obscenity, and profaneness were of concern to Richard Steele, controller of Drury Lane's Patent.[70] Anne Oldfield's performance as Susanna, along with her female form, came after only a few generations of actresses in the history of the English stage. A mere nineteen years after Jeremy Collier's *A Short View on the Immorality and Profaneness of the English Stage* and twenty years before the Licensing Act of 1737, *Three Hours after Marriage* displayed plenty of vice on the stage, alongside Gay's particular style of Augustan ridicule.

Moreover, the play's moment in the history of medical technology informs its attitude toward sex. The role of the male surgeon, with forceps in hand, in childbirth increased, as did his power over the female anatomy. The earlier anxieties eased as the new technology facilitated greater control of the female body. As Lisa Forman Cody remarks, "Reproduction intrigued the eighteenth-century public in ways not seen in the seventeenth century, and in so doing became an area of knowledge available to all, attainable through 'rational-critical' means rather than personal bodily experience."[71] The obvious irony is that *Doctor* Fossile is ignorant of contemporary advancements; he is thus unable to assess his own conjugal status. Three hours into his marriage, he foolishly and ignorantly tests his wife's virginity. He does not know her in the biblical sense but tries to know her with his mind; he fails.

As Katherine Mannheimer shows, Fossile's testing of his wife fits into a larger pattern, his obsession with textuality: "Fossile longs for a stably physical sign by which to know of a momentary event ('defloration') or even attitude ('lewdness,' or, later, 'shame' 2.470)—to convert these ephemeral comportments into stable textual emblems."[72] His fundamental nature is one of "sterility and petrification, . . . fossilization."[73] Coppola's conclusion is that Fossile is not an eighteenth-century man; he is an old-fashioned Galenic, concerned with humoral compatibility in his and Susanna's procreative potential.[74] This knowledge may be specious, but it is deployed to control Susanna nonetheless: whereas *The Changeling*'s treatment of women was violent, in *Three Hours* it is mercenary. An additional example of this is Fossile's niece Phoebe Clinket, an unsuccessful playwright and the other figure of virginal obsession. After several scandalous allusions to her "child," she clarifies that she is referring to her writing. As Susan Gubar points out, the Augustan satirist maligned female writers "as failures because they cannot transcend their bodily limitations; they cannot conceive of themselves in any but reproductive terms."[75] That Clinket's offspring is merely intellectual and her motherhood mere affectation is yet one more joke on the subject at hand; her body, as well as Susanna's, exists solely in patriarchal structures.

Shortly before Fossile administers the elixir to his wife, his actions repeatedly contribute to her commodification. He instructs his servant that both Susanna and "her Trunks and Boxes be immediately brought hither," intertwining her fate with that of her possessions in an almost metonymic relationship.[76] Next, he dwells on the anxious life that a cuckold is doomed to lead: "Ah *Fossile*! if the Cares of two Hours of a married Life have so reduc'd thee, how long can'st thou hold out! To watch a Wife all Day, and have her wake thee all Night! Twill never do."[77] Time is counted out and assessed. Wasted time is an undue burden on the obligatorily watchful husband. The commodification does not cease after Susanna has been cleared of unchastity by the dubious touchstone. Fossile absurdly invents a long-standing family tradition for the test: "These Drops have been a Secret in our Family for many Years. They are call'd the *Touch-stone of Virginity*. The Males administer it to the Brides on their Wedding-Day; and by its Virtue have ascertain'd the Honour of the *Fossiles* from Generation to Generation. There are Family Customs, which it is almost impious to neglect. . . . My Mother obliged me to this Experiment with her dying Words."[78] Fossile's embellishments reveal the importance of the female body: verification of chastity, passed down through the generations, would create a patriarchal pattern of ownership like any deed or legacy.

This is not the end of Fossile's scheming, though. Immediately after he is satisfied with Susanna's virginity, he expands his plot: "And to preserve her so, 'tis necessary that I have none but chaste Servants about her. I'll make the Experiment on all my Female Domesticks." This, addressed as an aside, provides the dramatic irony as he tells Susanna aloud, "I will now, my Dear, in thy Presence, put all my Family to the Trial. Here! bid my Niece, and all the Maid-Servants come before me."[79] The commodification of the female body extends to the lower-class servants around them. Naturally, the maidservants refuse to implicate themselves: they offer a variety of excuses and flee. The lone woman who takes the drink is Clinket, whose only offspring is her script and whose only love is Platonic, setting up another bawdy joke. As E. Parker, philomath, comments,

> *Fossile*'s putting his Family to the Trial with *Lubomirski*'s Drops, is a wretched Absurdity, here is a Stage full of People, who have no Business in the Play, brought in by Head and Shoulders, only for the introducing of this charming Definition.
>
> *Clink. How do the* Platonists *and* Cartesians *differ Uncle*?
>
> Foss. *The* Platonists *are for Ideas, the* Cartesians *for Matter and Motion.*
>
> This Vile Obscenity is call'd Humour; but more of this her after. The *Hoop-Petticoat* at the Close of this Act, is but a poor Incident, tho' the only one in the whole Farce.[80]

Parker's criticisms return attention to the sexual nature of Gay's farce. Both the double entendre and the petticoat scheme alluded to here focus audience attention once more on sex and the female anatomy. There is serious business between those lines and beneath that petticoat, however, as laughter disguises pragmatic concerns.

Sara D. Luttfring's concerns from *The Changeling* hold true here as well, with virginity "crucial in determining a woman's value on the marriage market and ensuring the legitimacy of her husband's familial line."[81] In the later period, Mary Fissell argues, women's reproductive bodies became "productive yet troubling analogies to the mysterious generativity of capitalism itself," and "older workshop metaphors about reproduction were supplanted by commercial and mercantile images."[82] Fissell finds this in the character of Polly Peachum in Gay's *The Beggar's Opera*: "the ways that female reproductive bodies are like the mysteries of the stock market are such a commonplace that they serve to make the audience laugh."[83] The metaphor of Clinket's material production (her play) as offspring is paralleled by the words used to describe Susanna's actual child, present onstage, in the final scene: "Invoice" and "Cargo."[84] The sailor who brings the infant on tells Fossile that "there is no Boarding a pretty Wench, without Charges one way or other; you are a Doctor, Master, and have no Surgeons Bills to pay; and so can the better afford it," and then he explains, "I was neither by when his Mother was Stow'd, nor when she was Unladen; Whether he belong to a fair Trader or be Run Goods, I cannot tell."[85] The mercantile and scientific concerns finally unite as Fossile is bested by the five proofs of filiation provided by a fellow doctor. Even though Fossile has not "had Carnal Knowledge of any Woman," he is declared to be the father of the child; and even as he protests his wife's virginity, the Lord Chief Justice's "Business" reveals Susanna's prior marriage.[86] Alone onstage with his child who is not his, Fossile delivers a final speech: "Fossile thou didst want Posterity: Here behold thou hast it. A Wife thou didst not want; Thou hast none. But thou art caressing a Child that is not thy own. What then? A Thousand, and a Thousand Husbands are doing the same Thing this very Instant; and the Knowledge of Truth is desirable, and makes thy Case the better."[87] Here, at last, anatomical truth is known, not through science but through the exchanges of sailors. This knowledge bests Fossile's "Carnal Knowledge" (or lack thereof); he prefers his "Posterity."

Harol points out that Fossile's language in the play "is marked by a wistfully excessive desire for scientific evidence of virginity, by skepticism of the medical profession's ability to provide such proofs, by a lament for the incompatibility of knowledge and sexual pleasure, and by a fear that the deceptive and inscrutable female body threatens masculine dominion."[88] While it is accurate to say that Fossile's problems arise "from his reliance on the methods of science," it is also scientific knowledge that allows the audience to be in on the joke.[89] Only an ignorant audience member would assume that a trick from a hundred-year-old tragedy would still work. Even though it is clear that Fossile is ignorant, a comic ending restores (patriarchal) order: in his final monologue, he gains what he desires (knowledge, posterity), while confessing that sex is unimportant. Though Fossile is a figure of satire, as John Fuller states, "our sympathies are surely with him."[90] Once succession has been secured, the technology of virginity is irrelevant.

One telling anecdote from the production history of *Three Hours after Marriage* contributes to our understanding of women's lost agency. On the fourth night of its original run, the following incident occurred: "The droll facetious [William] Penkethman, was that amphibious devourer, the crocodile, where the painter, the tailor, with other artificers had us'd their utmost skill. . . . Penkethman, the crocodile, boasting much in the beauty of his long tail, and, traversing the stage, unfortunately made such a parade with it, that he threw down Sarsnet (the attendant and confidant of Mrs. Townley) flat upon her back, where she discovered more linnen than other habiliments, and more skin and flesh than linnen; this began the first uproar in the audience."[91] Half an hour of chaos ensued. Adding insult to the injury of poor Sarsnet, the letter in which this anecdote appears concludes that "it is some danger, to give a woman room to talk."[92] The pasteboard creations were so popular that Mummy and Crocodile appeared in another Drury Lane farce decades later. The phallic shenanigans of male characters outlived any contribution from the women involved.

By the time of the late-century obstetric atlases by the renowned man-midwife William Smellie (1754) and William Hunter (1774), the level of anatomical detail had reached a point of unprecedented verisimilitude. Lyle Massey concludes that it is this scientific knowledge that brings "the pregnant female body and childbirth into the purview of the (male) medical specialist."[93] Paradoxically, though, the predominance of the medical aspects of reproduction, the forceps, the increase of surgeons regularly attending births, and the large number of male midwives led to a decreased ability to control women's sexual lives outside of pregnancy, including their virginity. Harol recounts that while "midwives previously had been called in by courts to confirm pregnancy, virginity, and even impotence, this function of the traditional midwife disappeared."[94] Despite science and medicine providing a greater understanding of female bodies to men, gaps were still present in that knowledge. These gaps were navigable, but only within a limited range.

Couched within the satire of *Three Hours after Marriage*, Susanna Townley's performative virginity functions on several levels. The first and most obvious is the mockery of an idiotic husband in a play that, as Winton says, follows broad comedic outlines: "A great deal of ink has been spilled in identifying the satiric victims in *Three Hours after Marriage*, much of it spilled in vain. The play is understandable without footnotes."[95] Susanna makes a fool out of her husband, passing his test, while her sexual history causes her no harm. Any potential progressive reading is undercut, however, by the realities of the eighteenth century. An economic reading of Susanna's performed virginity is more complex. While ninety-five years' worth of changes in science and technology transformed the virginity test from a tragedy to a farce, these changes came at the expense of the female body: Townley's sexually free existence is that of a mere exchangeable set of goods. Though *Three Hours after Marriage* pokes fun at Dr. Fossile, the women fare even worse. In the end, Susanna is trapped in her marriage to the unseen Lieutenant; after this revelation, she is silenced and never speaks again. Clinket, the laughingstock bluestocking, is unproductive in both mind and body. Her only hope would be to write a comedy based on these events, but that seems unlikely to be any good. Meanwhile, Fossile, though chastened, gets what he wants; patriarchal systems of control exert dominance. While scientific knowledge and technological advancement come with gains, they do not apply equally to everyone.

NOTES

Special thanks to Kaylee Pomelow, my research assistant.

1. E. Parker, *A Complete KEY to the New Farce, Call'd "Three Hours after Marriage"* (London, 1717), 9.

2. For summaries of existing scholarship, see Bruce Boehrer, "Alsemero's Closet: Privacy and Interiority in *The Changeling*," *Journal of English and Germanic Philology* 96, no. 3 (July 1997): 349–350; Dale B. J. Randall, "Some Observations on the Theme of Chastity in *The Changeling*," *English Literary Renaissance* 14, no. 3 (Autumn 1984): 353.

3. Randall, "Some Observations," 360.

4. Boehrer, "Alsemero's Closet," 353–354.

5. Randall, "Some Observations," 358–359.

6. George Walton Williams, introduction to *The Changeling*, by Thomas Middleton and William Rowley, ed. George Walton Williams (Lincoln: University of Nebraska Press, 1966), xiii; Martin W. Sampson, ed., *Thomas Middleton* (New York: American Book Company, 1915), 402n342.

7. For a survey of various tests and tricks, see Kathleen Coyne Kelly, *Performing Virginity and Testing Chastity in the Middle Ages* (London: Routledge, 2000), 28–33; Randall, "Some Observations," 355–358.

8. Johann Jacob Wecker, *Eighteen Books of the Secrets of Art & Nature* (London, 1660), 104.

9. Marie H. Loughlin, *Hymeneutics: Interpreting Virginity on the Early Modern Stage* (Lewisburg, PA: Bucknell University Press, 1997), 29–47.

10. Swapan Chakravorty, *Society and Politics in the Plays of Thomas Middleton* (Oxford: Clarendon Press, 1996), 159.

11. Thomas Laqueur, *Making Sex: Body and Gender from the Greeks to Freud* (Cambridge, MA: Harvard University Press, 1990).

12. Karen Harvey, "The Century of Sex? Gender, Bodies, and Sexuality in the Long Eighteenth Century," *The Historical Journal* 45, no. 4 (December 2002): 899–916.

13. Mary E. Fissell, *Vernacular Bodies: The Politics of Reproduction in Early Modern England* (Oxford: Oxford University Press, 2004), 12.

14. Ibid., 114.

15. Ibid., 165.

16. Lauren Kassell, "Medical Understandings of the Body, *c.* 1500–1750," in *The Routledge History of Sex and the Body*, ed. Sarah Toulalan and Kate Fisher (London: Routledge, 2013), 66.

17. Ibid.

18. Helkiah Crooke, *Mikrokosmographia* ([London], 1615), 235.

19. Nicholas Culpeper, *A Directory for Midwives* (London, 1651), 28–29.

20. Giovanni Benedetto Sinibaldi, *Rare Verities: The Cabinet of Venus Unlocked and Her Secrets Laid Open* (London, 1658), 57–58.

21. Ibid., 59–60.

22. Corinne Harol, *Enlightened Virginity in Eighteenth-Century Literature* (New York: Macmillan, 2006), 59.

23. Thomas Chamberlayne, *The Compleat Midvvife's Practice Enlarged* ([London,] 1659), 303–304; Jane Sharp, *The Midwives Book* (London, 1671), 266–267.

24. Caroline Bicks, "Stones like Women's Paps: Revising Gender in Jane Sharp's Midwives Book," *Journal for Early Modern Cultural Studies* 7, no. 2 (Fall–Winter 2007): 1–27.

25. Thomas Gibson, *The Anatomy of Humane Bodies Epitomized* (London, 1682), 147–148.

26. Ibid., 154.

27. *Aristotele's Master-Piece* (London, 1684), 93.

28. Ibid., 94.

29. Robert Barret, *A Companion for Midwives, Child-Bearing Women, and Nurses* (London, 1699), 50.

30. William Cowper, *The Anatomy of Humane Bodies* (London, 1698), table 51, fig. 3.

31. Michael Etmullerus, *Etmullerus Abridg'd* (London, 1703), 582–583.

32. Nicolas Venette, *The Mysteries of Conjugal Love Reveal'd*, 2nd ed. (London, 1707), 67–68.

33. Ibid., 69.

34. John Marten, *Gonosologium Novum* (London, 1709), 74, 75.

35. Nicholas Culpeper, *A Directory for Midwives*, rev. ed. (London, 1716), 231.

36. Ibid., 29.

37. Ibid., 232.

38. Ibid., 233.

39. Fissell, *Vernacular Bodies*, 196.

40. *Aristotele's Master-Piece*, 95.

41. Kassell, "Medical Understandings," 57.

42. Michael Stolberg, "Examining the Body, *c.* 1500–1750," in Toulalan and Fisher, *The Routledge History of Sex and the Body*, 91.

43. Kelly, *Performing Virginity*, 28–33.

44. John Leake, *Introduction to the Theory and Practice of Midwifery* (London, 1787), 5; emphasis in original.

45. Adrian Wilson, *The Making of Man-Midwifery: Childbirth in England, 1660–1770* (Cambridge, MA: Harvard University Press, 1995), 38.

46. Lisa Forman Cody, "The Politics of Reproduction: From Midwives' Alternative Public Sphere to the Public Spectacle of Man-Midwifery," *Eighteenth-Century Studies* 32, no. 4 (Summer 1999): 485.

47. Ibid., 486–487, among others.

48. Caroline Bicks, *Midwiving Subjects in Shakespeare's England* (Aldershot, UK: Ashgate, 2003), 62.

49. Roy Porter and Lesley Hall, *The Facts of Life: The Creation of Sexual Knowledge in Britain, 1650–1950* (New Haven, CT: Yale University Press, 1995), 15.

50. Tassie Gwilliam, "Female Fraud: Counterfeit Maidenheads in the Eighteenth Century," *Journal of the History of Sexuality* 6, no. 4 (April 1996): 518.

51. Harol, *Enlightened Virginity*, 69.

52. Harvey Graham, *Eternal Eve* (London: William Heinemann Medical Books, 1950), 188.

53. While it is easy to overstate the forceps' importance in the displacement of midwives, it may have been a necessary "precondition." Wilson, *The Making of Man-Midwifery*, 101.

54. Shannon Westin, "The Obstetric Forceps: Shrouded in Mystery," *Obstetrical and Gynecological Survey* 61, no. 6 (June 2006): 357.

55. Ibid.

56. Ibid.

57. Leake, *Midwifery*, cover, 43.

58. Bettina Wahrig, "Clocks with Hands: Instruments, Hands and Parturients in a Changing Horizon of Time," *History and Philosophy of the Life Sciences* 35, no. 1 (2013): 62.

59. Lyle Massey, "Pregnancy and Pathology: Picturing Childbirth in Eighteenth-Century Obstetric Atlases," *The Art Bulletin* 87, no. 1 (March 2005): 73–91; Cody, "The Politics of Reproduction"; among others.

60. Wilson, *The Making of Man-Midwifery*, 91–101.

61. J. L. Austin's concept of the performative (an utterance that "is the performing of an action—it is not normally thought of as just saying something") allows Judith Butler to ask performative questions of gender and the body: "To what extent does the body come into being in and through the mark(s) of gender? How do we reconceive the body no longer as a passive medium or instrument awaiting the enlivening capacity of a distinctly immaterial will?" Austin, *How to Do Things with Words* (Cambridge, MA: Harvard University Press, 1962), 7; Butler, *Gender Trouble*, 10th anniversary ed. (London: Taylor and Francis, 1999), 13.

62. Mara Amster, "Frances Howard and Middleton and Rowley's *The Changeling*: Trials, Tests, and the Legibility of the Virgin Body," in *The Single Woman in Medieval and Early Modern England: Her Life and Representation*, ed. Laurel Amtower and Dorothea Kehler (Tempe: Arizona Center for Medieval and Renaissance Studies, 2003), 232.

63. Sara D. Luttfring, "Bodily Narratives and the Politics of Virginity in *The Changeling* and the Essex Divorce," *Renaissance Drama* 39, no. 1 (2011): 98. Moving forward to studies of the eighteenth century, Soile Ylivuori cites Erving Goffman in saying, "chastity should be understood as a complex system of performances and presentations." Ylivuori even explicitly labels chastity as "performative" (although without citing Austin or Butler). Ylivuori, "Rethinking Female Chastity and Gentlewomen's Honour in Eighteenth-Century England," *The Historical Journal* 59, no. 1 (March 2016): 73, 74. Further, Sarah Salih applies Butler's ideas of performative gender to the earlier *Katherine* group, in which she finds the concept of virginity to be "a successful rearticulation of the heterosexual hegemony." Salih, "Performing Virginity: Sex and Violence in the *Katherine* Group," in *Constructions of Widowhood and Virginity in the Middle Ages*, ed. Cindy L. Carlson and Angela Jane Weisl (New York: St. Martin's Press, 1999), 98.

64. Bert O. States, *Great Reckonings in Little Rooms: On the Phenomenology of Theater* (Berkeley: University of California Press, 1985), 20.

65. Jill Dolan, *The Feminist Spectator as Critic* (Ann Arbor, MI: UMI Research Press, 1988), 57. See also Marc Silverstein's exegesis of Hélène Cixous's work, which "points to the coercive model of identification that characterizes the illusionistic stage—a model forcing women to define their subject position as spectators in terms of the victimization and objectification

demanded both by the unfolding of the narrative enacted on stage, and by a scopic economy that splits visual pleasure between an active male specular subject . . . and the passive female specular object." Silverstein, "'Body-Presence': Cixous's Phenomenology of Theater," *Theatre Journal* 43, no. 4 (December 1991): 507.

66. Fissell, *Vernacular Bodies*, 153.

67. Calhoun Winton, *John Gay and the London Theatre* (Lexington: University Press of Kentucky, 1993), 51; George Sherburn, "The Fortunes and Misfortunes of *Three Hours after Marriage*," *Modern Philology* 24, no. 1 (August 1926): 93; among others.

68. Al Coppola, *The Theater of Experiment: Staging Natural Philosophy in Eighteenth-Century Britain* (Oxford: Oxford University Press, 2016), 107.

69. Parker, *A Complete Key*, 9.

70. Winton, *John Gay*, 55. For an opposing view, see Sherburn, "Fortunes and Misfortunes."

71. Cody, "The Politics of Reproduction," 483.

72. Katherine Mannheimer, "The Scriblerian Stage and Page: *Three Hours after Marriage*, Pope's 'Minor' Poems, and the Problem of Genre-History," *Comparative Drama* 43, no. 1 (Spring 2009): 66.

73. Ibid., 70.

74. While the play satirizes its antiquarian, these views were not necessarily progressive. Coppola, *The Theater of Experiment*, 107–109.

75. Susan Gubar, "The Female Monster in Augustan Satire," *Signs* 3, no. 2 (Winter 1977): 389.

76. John Gay, *Three Hours after Marriage* (1717), in *John Gay: Dramatic Works*, ed. John Fuller (Oxford: Clarendon Press, 1983), 241.

77. Ibid., 241.

78. Ibid., 242.

79. Ibid.

80. Parker, *A Complete Key*, 10.

81. Luttfring, "Bodily Narratives," 98.

82. Mary Fissell, "Remaking the Maternal Body in England, 1680–1730," *Journal of the History of Sexuality* 26, no. 1 (January 2017): 115, 118.

83. Ibid., 138.

84. Gay, *Three Hours*, 255–256.

85. Ibid., 256–267.

86. Ibid., 260.

87. Ibid., 261.

88. Harol, *Enlightened Virginity*, 131.

89. Ibid., 133.

90. John Fuller, introduction to *John Gay: Dramatic Works*, ed. John Fuller (Oxford: Clarendon Press, 1983), 31.

91. Letter reproduced in *Three Hours after Marriage* (Dublin, 1761), 221.

92. Ibid., 222.

93. Massey, "Pregnancy and Pathology," 74.

94. Harol, *Enlightened Virginity*, 76

95. Winton, *John Gay*, 53.

CHAPTER 4

Gulliver's Travels, Automation, and the Reckoning Author

Zachary M. Mann

Lemuel Gulliver's "Voyage to Laputa, Balnibarbi . . . ," in the third book of what is commonly referred to as *Gulliver's Travels* (1726), is one of the most famous encounters between scientific knowledge and the literary imagination. For that reason, most scholarship on the third voyage has focused on the ways in which Jonathan Swift draws inspiration from and responds to the natural philosophers on his bookshelves. Scholars point to similarities between Swift's Academy of Projectors at Lagado and the Royal Society of London for Improving Natural Knowledge, founded in 1660.[1] Marjorie Nicolson and Nora Mohler noted that the impractical projects at Lagado resemble actual ones reported in the *Philosophical Transactions of the Royal Society*, and many scholars since have pointed out that Swift was well versed in the works produced by the Society, including Robert Hooke's *Micrographia* (1665), Thomas Sprat's *History of the Royal-Society* (1667), and *The Philosophical Works of the Honourable Robert Boyle* (1725). Swift even visited the Society at Gresham College in 1710.[2] The common refrain throughout Swift scholarship is that the author used this insight to shine a light on the cost of the Society's questionable priorities in the face of imperial brutality. The third voyage is often framed as a satirical response to these real-world projectors.

The experiment at Lagado referenced most often is the "Project for improving speculative Knowledge by practical and mechanical Operators," a machine powered by humans turning cranks that, by randomization and without human thought, "may write Books in Philosophy, Poetry, Politicks, Law, Mathematicks and Theology."[3] Scholars have discussed this "wonderful Machine," as Gulliver refers to it, in one of three ways: as a critique of nonexperiential knowledge making, traced often to Isaac Newton's aside in *Principia*, that "we could derive the other phenomena of nature from mechanical principles by the same kind of reasoning";[4] relatedly, as an early example of machine thinking that foreshadows the rise of the algorithm in digital culture; or, most often, as a parody of a burgeoning, unregulated print industry that figuratively produces nonsense. In every case, Swift is represented as an "ancient" responding to "modern" systems of knowledge dissemination popularized by the Royal Society (and a book trade that rewards its imitation). The Society operates on the notion that truth is discovered or generated rather than evident or abstract, and for ancients like Swift, this modern approach splintered truth into a heterogeneity of pseudo-truths produced either by experiments that transform idiosyncratic perception into fact or by misguided beliefs that just because it is printed, it must be true.[5]

Recently, scholars of science and literature have complicated the notion that Swift merely responds to the cultural production of the Royal Society. In *The Experimental Imagination*, Tita Chico argues that, rather than writing literary reflection alone, Swift participates in the shaping of scientific knowledge.[6] Writing on the second voyage, Gregory Lynall in *Swift and Science* compares Swift's matter-of-fact style to Hooke's, suggesting that Swift parodies and thus reveals "the scientific treatise as a literary form."[7] Gulliver's descriptions of the Academy experiments are not dissimilar, on the surface, to the *Philosophical Transactions*, right down to the inclusion of diagrams. And, as a print commodity, *Gulliver's Travels* resembled the other publications in circulation by Swift's publisher, Benjamin Motte, which tended to be works of natural philosophy.[8] Thus, Swift writes amid and in the form of modern knowledge dissemination, arguably contributing to its living discourse. One strategy of like-minded critics has been to put into conversation Swift's descriptions of the miniature Gulliver in Brobdingnag with later thinkers who conceive of the body and world as machine-like.[9] In such works, the popular interest in automata demonstrates a paradigm shift among the middle classes, and Swift's rendering of Gulliver as an automaton helps orient readers to this new worldview. Comparing the second voyage to future works, such as Julien Offray de La Mettrie's *Man a Machine* (1747) or the theatrical automata of Jacques de Vaucanson (1737), has elicited discussions of Swift as an agent in man's changing relationship to technology.

In this essay, I am also interested in Swift's position toward and contribution to the sciences of the eighteenth century. However, rather than focusing on Swift's relationship to figures such as Boyle and Newton, I am interested in the *things* that make up arguably the most consequential technology of his century: the machines found in textile mills. As Jennifer L. Lieberman points out in "Finding a Place for Technology," literary scholars tend to favor the discursive similarities between literature and science.[10] In so doing, they have not only left the material realities of *Gulliver's Travels* largely unexcavated but also neglected the tension caused by Swift's interest in both science *and* industry. In the century between the founding of the Royal Society and the Society of Civil Engineers (1771), machines moved from laboratories to mills, and narratives of technological innovation shifted from philosopher-inventors to industrialists like James Hargreaves and Richard Arkwright. Over the course of Swift's career, the British Empire was transitioning from its age of science to an age of machinery, and in his surviving correspondence, it is more common to find the author discussing woolen manufacture than the treatises of the intellectual elite. Swift, for whom the intervention of learned virtuosi into economic affairs is a highly contested and politicized arena, was intimately aware of—and even helped to shape—these flows of *techne*.

Attending to science's industrialization, in this essay I examine anew Gulliver's third voyage, troubling the Royal Society's privileged status in its scholarship and situating Swift's satire in histories of technology—especially the inventions of textile machines and their influence on constructions of labor. Ultimately, I reconsider the "Project for improving speculative Knowledge" in order to revisit Swift's critiques of knowledge making as they are embodied in a machine inspired by the inventions of its day. In fact, as the next section of this essay demonstrates, machine innovation figures throughout *Gulliver's Travels*. Rather than comparing the experiments at Lagado to works such as *Man a Machine*, I compare Swift's particular notions of automation to actual engines in development in Europe, from spinning mills to the automated silk drawloom and punch-card programming systems invented by Basile Bouchon in 1725, improved by Jacques de Falcon in 1728 and Vaucanson in 1745, and eventually perfected by Joseph-Marie Jacquard.

These drawlooms are uniquely situated in the history of technology because they span the philosopher-inventor's workshop to the textile mill. They embody the shift, which Francis Bacon intended for learned societies, from the experimental laboratory to the administration of public good. And they illustrate, for Simon Schaffer and others, that automata were not only metaphors but also "played a significant role in the manufacturing economy."[11] In this essay, I follow the work of Barbara Hahn, treating machines as nodes via which historians can connect science to histories of labor and commodities—even before the patents were issued. As Gillian Cookson writes in *The Age of Machinery*, challenges in wool and silk manufacture led to local innovations that predate Hargreaves's spinning jenny and Arkwright's water frame, only becoming part of the "Industrial Revolution" when industrialists scaled machines to meet the demands of the cotton trade.[12] The same can be said of Bouchon's and Falcon's drawlooms, which, due to reactionary views within the Lyonnaise silk industry, were tabled until Jacquard's lifetime. These early machines operated as harbingers of future mechanization, circulated in print by the intellectual elite. Their conceptual designs developed into engines of commerce via the same discourses within which Swift operated.

Reading the "Project for improving speculative Knowledge" through this material history reconceptualizes eighteenth-century knowledge production in industrial terms. Textile manufacture served as vanguard for all machine innovation; any changing attitudes toward labor in the textile trade (e.g., weaving) influenced analogous attitudes in other industries (e.g., book publication). Automated weaving technology promised to replace the craftsperson's role, even the act of composition, with a mechanical process; meanwhile, recent copyright laws inaugurated authors and inventors as autonomous owners of their own ideas. As Joseph Loewenstein writes in *The Author's Due*, any "transformation in the available means of storing and distributing information inevitably requires of a culture that it renegotiate its conception of how information is *produced*."[13] While new technologies were revolutionizing the concept of material production in England, legal discourses—from the Statute of Monopolies to the Statute of Anne—were engaging in the individuation of creative endeavor centered on the human subject.[14] Swift's "wonderful Machine" thus provides an opportunity to discuss the methods by which the figure of the author or inventor is separated, conceptually, from more mechanical aspects of intellectual labor at a moment when these categories are in flux. As this essay shows, despite Swift's sympathies for the peoples below Laputa, his work contributes to the aforementioned cultural project by using machines of early industrial innovation to hierarchize types of thinking, constructing the industrial half of authorship as mere reckoning.

Spinners and Weavers

Throughout *Gulliver's Travels*, the textile industry recurs in three ways: as an anchor for Gulliver's unstable relationship to the size of objects, as a narrative strategy to describe imagined worlds in familiar terms, and as an allegory for England's economic oppression of Ireland. In regard to the first, consider how the Lilliputians describe Gulliver's pocket watch: "he put this Engine to our Ears, which made an incessant Noise like that of a Water-Mill." Here clockwork, emblematic of the small, intricate designs of seventeenth-century virtuosi, are translated by Swift's shifting positionality into the crude batteries of more recent industrial production. In another episode, Gulliver makes dainty Lilliputian artifacts of war by comparing them to knitting needles; in multiple other chapters, the fineness or coarseness of clothing reflects his size in relation to the livestock of his hosts. And in one especially memorable image from the second voyage, a miniature Gulliver compares the purring of a cat to "a noise behind [him] like that of a dozen Stocking-Weavers at work."[15] A dozen machines in operation suggests something like a mill; a stocking-weaver is "one who weaves with a stocking-frame," the mechanical knitting machine invented by William Lee in Nottingham that kick-started the first Industrial Revolution.[16] In this view, the Brobdingnagian farmer's cat audibly signals the transition of textile production from handwork, something small rural families might have toiled at on their own, to mechanized mass production, emasculating the diminished stature of a man—figuratively for the farmers and literally for Gulliver. Thus, Swift's horrors of scale can also be read as anxieties toward the proliferation and increasing size of industrial artifacts.

As Jonathan Sawday makes clear in *Engines of the Imagination*, Europe at the time was not lacking for machine noise; it would not have been strange for Swift to compare a loud sound to mill power.[17] Stocking-frames, by then in use for over a century, would have been easy to recognize for a general readership, and stocking-weavers had over that time become consolidated under single roofs. Hahn writes anecdotally that, by the 1720s, "one hosier employed more than forty apprentices" and, to hide the machines, "erected a large building in Nottingham."[18] In general, as Cookson notes, "textile technologies, which evolved steadily over the centuries, saw a gathering change in pace in the early eighteenth century."[19] By 1717 there was a flax mill in Ireland and, by 1726, a lint mill in Scotland, both powered by water.

The most famous inventions were those that mechanized spinning—the process of making fibers into yarn—and while these culminated in Samuel Crompton's spinning mule of 1779, they belong to a long history of innovation that begins with patents in 1678.[20] Arkwright's 1769 frame, which used rollers to draw and twist fiber, was not much different from John Wyatt and Lewis Paul's 1738 patent, which was in turn adapted for wool and cotton from an Italian silk-throwing system that had already been imported to Derby in 1702.[21] In other words, that Brobdingnagian house cat was compared to increasingly normalized phenomena.

As others have pointed out, Swift probably borrowed such machines to render quotidian his most speculative inventions. For instance, in describing the "floating" technology of Laputa, Swift contrives "a Load-stone of a prodigious size, in shape resembling a Weavers Shuttle."[22] Lynall suggests that, "by domesticating this technology of the foreign 'other' through direct comparison with European invention, Gulliver's description resembles the rhetorical strategies of many essays published in the *Philosophical Transactions*."[23] What Lynall does not mention is that the inventor credited with the "flying" weaver's shuttle, John Kay, was not a member of the Royal Society but a hand-loom reed maker in Lancashire. The innovation was a piece of wood with which the weaver could, by pulling a string, send the shuttle "flying" along a track without the need of an assistant. The mythic version of the story dates Kay's patent to 1733 and credits him for creating the bottleneck necessary to inspire future innovations in spinning. However, revisions to this history suggest that the flying shuttle was both adopted very gradually and in development for years prior.[24] Though Swift is unlikely to have based his "Weavers Shuttle" on Kay's, he knowingly draws comparisons between a device by which the island of Laputa floats above Balnibarbi and an object very much at the center of ongoing innovations in weaving. Thus, Gulliver makes practical the wondrous, but he also translates that science into objects that directly impact the economies of the British Empire.

Swift was more than interested in the textile industry. In multiple letters written during the 1720s, he blames the economic crisis in Ireland on England's Navigation Acts and especially the Woollen Act of 1699, which relegated Ireland to a colony, prohibited the exportation of Irish woolen manufacture, and forced Ireland to export its raw wool to heavily levied English ports.[25] The cause was dear to him; he even met with Dublin's Corporation of Weavers in Woollen and Silks multiple times to encourage more sustainable practices in the Irish textile sector. (Despite Gulliver's seeming loyalty toward England, there was probably much bitterness behind Swift's words when Gulliver hopes the flock of Lilliputian sheep he delivers to England would "prove much to the Advantage of the Woollen Manufacture, by the Fineness of the Fleeces.")[26] As Lynall points out, in comparing the technology by which Laputa wields martial power over its colonies to a weaver's shuttle, Swift uses a piece of a loom "as a symbol of Britain's economic grip over Ireland." This was legible rhetoric; a "key" published alongside *Gulliver's Travels* identified the load-stone as "a just Emblem of the British Linen and Woollen Manufactures."[27] While some of Swift's ire is reserved for the projectors, his letters place greater blame on the industrialists in Yorkshire and Lancashire, whose innovations he resents for spinning Irish wool. There is something whimsical about a floating island, just as the many contrivances of learned societies seem harmless in their laboratories. But for Swift, when mechanized to an industrial scale, the most innocent ideas can incur terrible costs.

In those letters, Swift calls for the people of Ireland to start wearing only Irish wool and asks Irish weavers to improve their manufacture to further reduce demand for imported goods. One of the experiments in the Academy at Lagado seems to share Swift's sympathies. When Gulliver enters this room, he finds it covered in spiderwebs. Inside, an "Artist" explains his project to replace imported silkworms with "Domestick Insects."[28] The plan mirrors Swift's calls for support of Irish over foreign manufacture; here the spider provides the possibility of locally sustainable textile production while the Chinese or French silkworm stands in for English imports threatening the Irish economy. Scholarship tends to agree that Swift, in his representations of Laputan experiments, makes material what the *Philosophical Transactions* convey in aspirational language. But in this case, Swift literalizes a recurring reference in his own letters against the Navigation Acts: Ovid's fable of Arachne and Pallas. In "A proposal for the universal use of Irish manufacture," Swift writes of the goddess who, jealous of Arachne's weaving, "turned her into a spider, enjoining her to spin and weave for ever, out of her own bowels, and in a very narrow compass." Swift admits that he "always pitied poor Arachne, and could never heartily love the goddess on account of so cruel and unjust a sentence." (It should be noted that Ovid's Minerva attacks Arachne with a weaver's shuttle, similar to the way Laputa uses something "resembling a Weavers Shuttle" to rain cruelty on the lands below it.) The analogy paints England as the bitter goddess and Ireland as the poor spider, turning a parable about mortal hubris into one about retaining autonomy over one's labor.

When it comes to spinning silk, the silkworm was at the time the only viable resource. But Swift renders the Laputan Artist's spiders more able than cursed; unlike silkworms, they "understood how to weave as well as spin." Gulliver's version of Arachne turns the myth into techno-optimism, a liberating *techne* by which Ireland can potentially reverse its sentence. One reason the Artist's spiders "excelled" silkworms, it turns out, is that they are *programmable*. He explains to Gulliver that the color, strength, and texture of the silk can be determined ahead of time by what he feeds the flies. Gulliver observes "a vast number of Flies most beautifully coloured," and according to the Artist, "the Webs would take a Tincture from them," and "Food for the Flys of certain Gums, Oyls, and other glutinous Matter . . . give a Strength and Consistence to the Threads."[29] Thus, through a kind of insect husbandry, the customized weaving of silk is predetermined, requiring no additional human labor. While this project will probably suffer the same fate as the others at the Academy, for this momentary fantasy, Swift borrows something only recently possible in industrial sectors: the mechanical production of a commercial good that is repeatable through the use of objects inscribed with instructions by human hands. Since the drawloom first migrated to Italy in the thirteenth century, the race to invent the perfect "coded object," and thereby to be able to automate the weaving process, motivated much of Europe's machine innovation. Swift's "Flys" behave not unlike the cylinders invented by Vaucanson for his automata or, invented by Bouchon only a year before *Gulliver's Travels* was published, the perforated paper tape of the first programmable loom.

By 1725, Bouchon figured out a way to replace the communication between weavers and their assistants with perforated paper tape, wherein a simple binary—whether or not a hole in the paper blocked a spring-loaded needle—determined if the assistant should pull down on the lash that, in turn, lifted the warp (or group of threads).[30] Falcon soon after replaced this tape with a sewn-together chain of hole-punched pasteboard cards—or "punch cards." These coded objects, the most important such objects in machine programming history, essentially removed intellectual labor from the production of patterned silk. Even the weaver's labor was minimized, for all patterning was done beforehand and could be reproduced as long as the punch cards kept their integrity. Eventually Vaucanson designed a punch-card system to fully control the warp (and thus, by alerting the machine whether or not to lift the thread groups mechanically, rendered unnecessary the assistant altogether). By the nineteenth century, designers only needed to draw the preferred pattern and send it to an artist; the artist then punched holes in pasteboard cards such that, when fed through a jacquard loom, it would weave into the silk the exact pattern preconditioned by the manufacturer. With dyes, gums, and oils instead of a hole-puncher, Swift's Artist accomplishes the same with flies.

Swift's spiders do not predict the jacquard loom, but the spider room in the Academy at Laputa resembles a vision of automation. This vision would be articulated in the November 1745 edition of *Mercure de France*, prompted by Vaucanson's invention: "a machine with which a horse, an ox or an ass [can] make cloth more beautiful and much more perfect than the most skilled silk-workers."[31] Though Vaucanson's drawloom failed to change the industry, it would have, in theory, made it possible for the weaver to merely stand around as nonhuman forces did the work, like the Artist in Laputa, who only watches the spiders as they spin and weave without need for direction. Swift maps this imaginary onto tiny animals. Similar scenes occur in the first two voyages: a horde of Lilliputians—including "two hundred Sempstresses" and "three hundred Taylors"—crawl over Gulliver's giant body to measure him for clothes, reminiscent of the organized movement of ants; the aforementioned sheep that Gulliver delivers to England are bug-like in size; and the king of Brobdingnag alternatingly refers to the tiny Gulliver as "a piece of Clock-work" and an "insect."[32] With the exception of the purring cat, these otherwise human or mammalian figures are described as insectoid, even *verminous*. Some, including the cat, are also rendered machinic. The silkworm spins because it is genetically programmed to do so. The spider weaves in geometric patterns. Groups of insects move as if synced to a network. In a work associated with questions of human versus animal, mechanical labor forms a new (if still animalistic and subhuman) category.

The Knowledge Mill

Two of the more memorable machines in *Gulliver's Travels* are not related to textile production, at least at first glance. The first is designed so Gulliver can turn pages in Brobdingnag. The tiny Gulliver required this machine because the books were "as thick and stiff as a Past-board, and . . . eighteen or twenty Foot long." The second machine, oddly the same size as Brobdingnagian books, is the perversely bookish "Project for improving speculative Knowledge by practical and mechanical Operators."[33] In this other room at the Lagado Academy, Gulliver finds a "Professor" with "forty Pupils about him" and a twenty-foot "Frame." The frame holds "all the Words of their Language in their several Moods, Tenses, and Declensions, but without any order," glued to wooden blocks strung by wires. Thirty-six of the students turn cranks to randomize these blocks while the remaining four write down any legible sentences formed by happenstance, with the aim to produce "a compleat Body of all Arts and Sciences."[34] Using only human muscle power, the "engine" generates books consisting of randomized words. Swift here literalizes a strategy that he often uses to satirize shortcuts in writing; in fact, he probably also borrows from his own aside in "A Tritical Essay upon the Faculties of the Mind" (1707), that he would not believe "the accidental jumbling of Letters of the Alphabet would fall by chance into a most ingenious and learned Treatise of Philosophy."[35]

Many scholars interpret the engine as a satire of philosophers who have attempted to instrumentalize thought. As Joseph Drury observes in *Novel Machines*, natural philosophers engaged, broadly, in the "mechanization of knowledge making through the introduction of formal rules of method."[36] Lynall suggests that Swift is responding to calculating engines like those invented by Blaise Pascal (1642) and Gottfried Leibniz (1673), whose designs, "to perform mechanically all the operations of arithmetic with certainty and expedition," were regularly discussed in the *Philosophical Transactions*.[37] Leibniz envisioned, using mathematics as a universal language, the construction of an encyclopedia that would include all human knowledge and a calculating machine that could parse that knowledge using logic based on Aristotelian categories. Along these lines, Pamela McCorduck in *Machines Who Think* calls the engine "a dotty burlesque of Ramon Llull's *Ars Magna*." Llull's proposed "rings," articulated disks that recombined old maxims to form new insight, were a popular target; François Rabelais in *Gargantua and Pantagruel* and Bacon in the *Advancement of Learning* wrote off the "Art of *Lullius*" as "vanity and imposture" and "nothing of worth," respectively. Martin Gardner, in *Logic Machines and Diagrams*, even writes that the *Ars Magna* "amounted virtually to a satire of scholasticism."[38] Such critics point to Swift's ire toward these kinds of get-knowledge-quick schemes and anyone hoping to repair, as the Professor laments to Gulliver, "how laborious the usual Method is of *attaining* the Arts and Sciences."[39]

Swift returns to this subject multiple times. In *Gulliver's Travels*, a different Lagado experiment proposes the delivery of scripture to the brain via the consumption of "a thin Wafer," inscribed on in tiny lettering.[40] And in *A Tale of a Tub*, the author devises a step-by-step recipe for attaining truth by snorting the desired books.[41] The recurring metaphors of the digestive system are no coincidence; for Swift, what degrades the learned subject is precisely its translation into material functions, into machines with cogs that grind their teeth. In *The Battle of the Books* (1704), Swift imagines a "Modern Spider of Invention"—a creature who "displays his great Skill in Architecture, and Improvement in the Mathematicks" but, because he uses those principles to avoid learning, produces only cobwebs. This metaphorical spider, rather than attending to geniuses of the past through study, feeds on—or is, like the weaving spiders in the other room, *programmed* by—"the Insects and Vermin of the Age"; like "Modern" philosophers who only regurgitate their contemporaries, the spider only ever "Spins and Spits wholly from himself" to produce cobwebs "spun out of [his] own Entrails." Swift refers to this programmatic knowledge—borrowed from others and untested in reality—as "the Guts of Modern Brains."[42] The Laputan engine, which uses engineering to produce such "Guts" spun from its own cog work, is the tragic tale of Arachne written onto intellectual labor.

The Professor's engine does more than simply randomize; it also *automates* production of that so-called knowledge. The labor that cranking replaces is the act of composition, such that after the Professor "emptyed the whole Vocabulary into his Frame," the combining of those words does not require a thinking subject.[43] John Chalker attributes the engine's inspiration to John Peter's pamphlet *Artificial Versifying, or the Schoolboys Recreation: A New Way to Make Latin Verses* (1677).[44] Swift would have read the review in *The Spectator* (1711), which refers to Peter as "an ingenious Projector" who, with "Poetical Logarithms . . . divided into several Squares . . . inscribed with so many incoherent Words," is "able to compose or rather to erect *Latin* Verses."[45] Swift's engine also "erects"—that is, it builds, constructs, or assembles rather than writes—strings of random words. In these cases, the building blocks of literature are vocabulary alone; however, Swift elsewhere has discussed using prewritten phrases (or, to use a twentieth-century term, *lexia*) in the composition of literary works.[46] His friend Alexander Pope, as Martinus Scriblerus, imagines this in "A Project for the Advancement of the Bathos" (1727), which mocks the influence of the "other arts of this age" on literary production by proposing the collection of works by specialists—"ironists," "allegorists," "simile-makers," and the like—into a "cabinet" from which a poet or orator will draw out the phrases they need, like playing a piano.[47] Here Pope, the dominating literary figure at the time, attacks the trend in the book trade of programmatic thought; and one might say Swift's engine functions as a metaphor for this kind of recombinatory composition.

For Swift, such methods have real consequences, especially when used for knowledge making. Recently, historians of the Enlightenment have framed the era as a media event in which written words, through their proliferation in print, begin outpacing the ideas themselves; as Brad Pasanek and Chad Wellmon write in "The Enlightenment Index," this new literary production "points less to ideas, authors, philosophers, or social organizations than from one page to another."[48] Scholars since, this one included, create their own repositories of notes on "paper slips" (real or virtual), with which they compose new works and disseminate new knowledge via reorganization (or regurgitation).[49] In *A Tale of a Tub*, Swift bemoans those who collect quotations from compendiums rather than original works, thereby treating thoughts as commodities and replacing human memory with the objects of their inscription. Kelly Swartz, in "The Maxims of Swift's Psychological Fiction," writes that Swift structures *A Tale of a Tub* as "aphorism after aphorism," and his author becomes a mere "collector of cliché."[50] For Swift, this new "Method, to become *Scholars* and *Wits*, without the Fatigue of *Reading* or of *Thinking*," is a dangerous one—and thus the victim of both his "Modern Spider" and the Laputan knowledge machine.[51] After the Licensing Act lapsed in 1694, publication expanded, and standards dropped in favor of profits. The index became activated as a mechanism by which knowledge multiplies into all possible iterations. In Swift's view, a new generation of philosophers could, by recycling the intellectual labor of others via works of reference published or collected, concern themselves with self-promotion rather than the advancement of learning.

One might read Swift's engine, then, for its resemblance to what Marcus Krajewski in *Paper Machines* calls "the scholar's machine": the manipulation of card indexes and paper slips such that, by allowing the scholar to reorder previously written notes into something virtually new, it behaves as a "text generator." Jean Paul, in *Selections from the Devil's Papers* (1789), wrote (incorrectly) that "Gulliver saw a machine in Lagado that mixed paper slips to such an extent that anyone they were read to could not tell whether they were read from an ordinary book or not."[52] Paul's misreading of Swift's engine imagines this scholarly practice embodied in a wooden frame: each paper slip or index card is a wooden block randomized by the students' cranking. The frame, in this view, becomes the scholar. Or, to put it another way, these *lexia* become coded objects—like "Flys" or punch cards—manipulated by assistants in the automated manufacture of literature. In an illustration included in the second edition of *Gulliver's Travels*, the frame even resembles a loom. Due to Swift's interest in the textile industry, and considering that the illustrations of stocking-frames were in regular circulation, it is likely that the engine—and probably any mechanical system of wooden blocks, wires, and assistants—would be influenced by real-world machinery; according to the *Oxford English Dictionary*, the word "Frame" was commonly used to refer to machine knitting.[53] The comparison is not perfect; yet the engine more accurately *weaves* rather than *writes* its "Books in Philosophy, Poetry, Politicks, Law, Mathematicks and Theology."[54]

Swift elsewhere is quick to compare literary and textile production. In his "Letter to the Archbishop of Dublin," for instance, he adds that not only woolen manufacture but also literature is more fashionable in Ireland if it is considered "from England." And, in a letter to Motte, Swift encourages Irish booksellers to smuggle their Dublin-printed books to London, just as he encourages Irish wool manufacturers to export "to any Country in Europe . . . and conceal it from the Custom-house

Officers," as he would hide his "Purse from a Highwayman." Referring to a perceived ban on Irish-printed literature, Swift even compares the Statute of Anne to Navigation Acts.[55] As many historians of authorship have pointed out, the invention of copyright is also the transformation of ideas into commodities. Swift's concern with programmatic knowledge is influenced by his concerns regarding for-profit writing, too, for which textile goods are a useful derogatory metaphor. This is similar to the way Bacon compared Llull's methodized formation of knowledge to the secondhand clothing in "a Fripper's or Broker's shop." Thus, it is fitting that the aforementioned review of Peter's pamphlet further satirizes Peter's (and others') approach to poetry with mention of "the Project of a *Dutch* Mechanick, viz. a Mill to make Verses,"[56] and that Swift's Professor at Lagado refers to the "reasonable Charge" necessary to operate his engine. It is also fitting that the knowledge engine's books are ultimately the product of the "bodily Labour," "Six Hours a-day," of forty students.[57] While it is not the factory scale of "two hundred Sempstresses" and "three hundred Taylors," here Swift has imagined, in economic terms, a textile mill for the production of knowledge.[58]

By the end of the century, Swift's engine—often standing in for a fantasy that machines could automate literature—metastasized into a way for critics to devalue the unoriginality of literary works. One reviewer for the *Monthly Catalogue*, for instance, regarding the novel *The History of Miss Maria Barlowe*, accused a whole swathe of modern novels of being produced by "the famous machine of [Swift's] Balnibarbian."[59] In such cases, what was machine-like was the tendency for novels to borrow from formulas (e.g., the happy ending). But for Swift and his contemporaries, commercialism, not a lack of originality, is what earned the mechanical metaphor. Allison Muri writes in *The Enlightenment Cyborg* that "the rise of a merchant class had created a 'mechanical' identity out of the very tools of a man's trade."[60] Likewise, in "The Mechanical Operation of the Spirit," Swift puts down opportunistic writers of religious works by referring to them as "British Workmen" toiling at a "Trade." And, in *A Tale of a Tub*, he describes the modern critic as "a sort of mechanic, set up with a stock and tools for his trade, at as little expense as a tailor."[61] As authorship became its own trade, a way to make a living, comparing written works to the products of mechanical labor became a way to devalue them; Swift's critique of such authorship relegates it to a form of labor unworthy of his creative class. Like the imagined output of his engine, these works did not require intellect. They were not so different from the self-spun cobwebs of other machines.

Author-Inventors

When Gulliver parts with the Professor at Lagado, he promises to "do him Justice, as the sole Inventor of this wonderful Machine," and says "that he should have the Honour entire without a Rival."[62] Gulliver's choice of words, to identify the Professor as an "Inventor" rather than an author, and by comparing his status elsewhere to "the right Owner" of the entire "Honour" (analogous to a royal patent), places the Professor in the role of the industrialist. Gulliver says nothing about the "compleat Body" of works that the engine will supposedly produce, though it is implied that he would profit from them; the Professor is "the right Owner" of the frame's designs and thus its manufacture. Throughout the eighteenth century, debates for authors' rights to ownership piggybacked on patent law by arguing for "the Resemblance between a Book and any other Mechanical invention." As Mark Rose points out in *Authors and Owners*, often property produced by hand was confined to the physical object while the fruits of intellectual labor extended beyond manuscripts to the ideas within.[63] The printing of a book was one labor, but the thoughts put into that book were more highly valued. Due to this growing legal and philosophical divide between making and thinking, the labor of the mind became more venerated and mystified, culminating in "lone genius" figures like the late poet John Milton.[64] The more physical the production of writing is, the more the intellectual labor is located not in the composition but in the concept.

Thus, it is the idea of the machine alone that warrants the Professor Gulliver's praise. As Matthew L. Jones writes in *Reckoning with Matter*, the cultural project to insist on authors' rights "helped make invention more intellectualized" too: machines were credited not to the contribution of the artisans who built them, only to "the ideational designs of philosopher-inventors." Jones traces an evolution from the single figure of an artisan-inventor, getting their hands dirty, to "a bifurcation of design" in which one person came up with the idea and others made it material.[65] In the early seventeenth century, Pascal was one of the first to argue that "artisanal labor should be noncreative" and subordinate to the philosopher's vision. This opinion would be reaffirmed by the authors of the many encyclopedias published in the decades after, including Denis Diderot, who ignores the role of artisans altogether in the descriptions for his *Encyclopédie*, erasing their intellectual labor. Diderot relegates any nonphilosopher to a noninventive class that only "practice[s] the mechanical arts that need the least intelligence."[66] With Gulliver's promise, the Professor also participates in this tradition; despite needing the "bodily Labour" of his students to produce words, and despite the machine's fully automated method of composing those words, the Professor gets sole credit. Swift, literalizing this labor relationship via mechanical metaphor, deconstructs the legal category of "Inventor" as one that privileges ideation over industry regardless of who labored.

This design bifurcation stays true when there is no philosopher involved, which was often the case in textile history. Most tales of industrial innovation are as fictional as Swift's Professor, spun by Victorian-era historians as part of the invention of the Industrial Revolution. Before Cartwright patented his power loom, for instance, he "had never before turned [his] thoughts to any thing mechanical, either in theory or practice."[67] Before bringing his business savvy to spinning machines, Arkwright was a barber. Hargreaves was an illiterate handloom weaver, such that Robert C. Allen, in "The Industrial Revolution in Miniature," goes so far as to say his jennies "owed nothing to science."[68] Gravenor Henson's *History of the Framework-Knitters* (1831) describes many more such cases, and John Theophilus

Desaguliers admits that "Men quite illiterate have often produc'd wonderful Engines."[69] Yet the rest of the people involved in those inventions—artisans, engineers, and the like—become a footnote to the accomplishment of one man who is made to look, in hindsight, as the source of the ideas. To maintain this image of the "lone genius," all innovation is attributed to whoever can be constructed as the one who labors the *most* intellectually—or, perhaps more accurately, the *least* mechanically. This required a recursive effort to hierarchize the different types of labor. Swift, for his part, lampoons this process when the Laputan tailors who design Gulliver new clothes "despise [practical Geometry] as Vulgar and Mechanick."[70] Their self-fashioning as philosophers relies more on their abjection as artisan-class labor than on their philosophical output. Or, to put it another way, the Professor's labor is intellectual because he does not crank or turn.

In eighteenth-century conceptions of authors and inventors, then, there are two kinds of thinking. Real thinking is inventive, even creative, while mechanical "ratiocination"—to quote Thomas Hobbes—is reduced to mere "reckoning."[71] Thinking with the body, whether manual labor, instinct, or emotion, was a lower order of intellect, that which the liberal human subject is defined against. Thus, narratives of invention are often paired with emphases on the lack of sentience beyond the inventor. Diderot in his *Encyclopédie* notes that weavers often work for forty years in mills "without knowing anything about these machines" and that a stocking-frame "makes hundreds of stitches at once . . . without the worker who moves the machine understanding anything, knowing anything or even dreaming of it."[72] Inventors of automated literary production tend to similarly demote the mental capacity of would-be such "authors"; for instance, Peter advertised his pamphlet with the promise that anyone of the "meanest capacity" or "altogether ignorant of arithmetic, and of all literature" could still utilize his poetry generator.[73] The *Spectator* review burlesques this language, adding, "any one without knowing a Word of Grammar or Sense, may, to his great Comfort, be able to compose . . . *Latin* Verses."[74] Pope promises the same in his "A Receipt to Make an Epic Poem," that "epic poems may be made *without a genius*, nay without Learning or much Reading."[75] Swift echoes these comments when the Professor explains to Gulliver that "the most ignorant Person . . . without the least Assistance from Genius or Study" can use his engine to produce the greatest of all works.[76]

Swift elsewhere suggests that automated productions are "ever in greatest Perfection, when managed by Ignorance."[77] While he is referring to the "Art of Canting," the sentiment reflects that of Adam Ferguson decades later: "Many mechanical arts require no capacity. They succeed best under a total suppression of sentiment and reason."[78] Swift often sides with the people caught beneath the intellectual labors of natural philosophers and figures of state, but he also disapproved of the way the mechanical arts promised the less learned that they could be taken as seriously as the great minds of the past. His sympathy has a limit. This is because, as Muri writes, "to imagine intellectual labour as a commercial, manual, and mechanical labour constituted an alarming diminishment in the status and profits of the educated gentlemen."[79] Swift pokes fun at the Laputans for claiming that an artisan's craft is not intellectual enough to count as science, but he also puts down machine thinking at a time when the intellectual elite is treating textile workers as docile automatons in service to philosopher-inventors. His Laputan writing machine is a social critique of a kind, but he neglects the actual victims of that experiment: the students. Allen, discussing the empire's industrialization, points out that the rapid innovations during the eighteenth century happened in England because the high cost of craftspeople required the mechanization and thus optimization of the labor force, a process that Ferguson and then Marx responded to in a growing political discourse.[80] That process occurred both on the factory floor and in pamphlets, tracts, and fiction, to which Swift contributed, that helped shape humanity's changing relationship to new industrial technologies.

Textiles aside, the technologies closest to Swift were the printing presses that operated, like the author of *A Tale of a Tub*, "too freely." Beyond mechanical metaphors, Swift attacked London's commercial printers on Grub Street by borrowing the image that Pope popularized in *The Dunciad*, in which "grubstreet" becomes, as Johnson defines it in his *Dictionary*, "any mean production."[81] He was also not shy in using the street name's other meaning; a "grub," after all, was the larva of an insect. Critics over the years have suggested that the central question of *Gulliver's Travels* is "what sort of animal man, as a species, really is."[82] While many critics champion Swift's attacks against rationalism, others, like Muri, note that Swift belongs to an intellectual class whose "moral authority relied upon differentiating lower-order mechanic intellect from higher-order nervous intellect." While Swift was putting down the Laputans for their allergies to the "Vulgar," he was also putting down the "gross corporeality" of mechanical labor, not unlike how the Houyhnhnm treated the Yahoos.[83] His metaphors of vermin—grubs for hack writers, flies for coded objects, and spiders for spinning and weaving—associate early forms of machine thinking with a certain kind of animal: creepy, crawly reckoners. For Swift, these technologized animals are useful figures via which an authorial subject can abject mechanical or commercial processes from their intellectual labor. But while Swift's use of reckoning vermin as meme might be aimed at modern knowledge production, it also reveals (and contributes to) one strategy through which certain kinds of industrial labor are dehumanized.

When scholars of posthumanism look to this era, they discuss how technology was used to define "the human," who can innovate and invent, against "the animal," which cannot. The liberal human subject was (and is, as scholars have recently made quite clear) in a crisis—in a state of flux, redefined from moment to moment, and tied to the plasticization of others against which that subject is defined.[84] That "other" has alternately been defined as irrational, as uncivilized, even as untechnologized in cases where the use of informational technologies distances the educated subject from the manual laborer. At the same time, the *things* of technology threatened the autonomy of human thinking and, in the case of punch-card systems, their thinking entirely. As N. Katherine Hayles points out, machine encroachment into eighteenth-century knowledge formation also presented a new challenge: the idea that "the human" was becoming a posthuman subject, "a material-informational entity whose boundaries undergo continuous construction and reconstruction."[85] When mechanical inventions destabilized the distinctions between humans and nonhuman labor (as well as between humans and other humans), people in privileged positions used discursive practices like science and literature to render their own category the exception: in this case, by constructing some forms of humanity (i.e., the reckoning author) further along the spectrum toward animality, as menial and automatic.

In *Man a Machine*, La Mettrie writes, "Man was trained like an animal. . . . He became an author in the same way as he became a porter."[86] Perhaps instead, as England mechanized its labor force in the eighteenth century, he became an author the same way he became a stocking-weaver—subjected into rather than trained for an industry that was further becoming a machine itself. Efforts to define categories such as "the author" as separate from this technologization have historically demoted machines to the status of mere tools. Swift is doing something else; by privileging the tools of composition in his satires of modern knowledge making, Swift is forced to reclaim the purely human author-inventor by other means. One way he does so, it seems, is to borrow from the well-rehearsed method described earlier, to define human subjectivity, again, against not only nonhuman mammals or so-called uncivilized others but those tiny creatures whose behavior and low-level cognition resemble the automatic qualities of the century's proliferating machines. Thus, scholars must look past the contradictions of Swift's "rational man" and reconsider the eighteenth-century author as a figure embedded in a world where engines both spin wool from colonies and churn out poetry—who, like Swift, must occasionally take the spider's point of view.

NOTES

1. Swift includes in his satire all learned societies. This line of scholarship goes back to the eighteenth century; the Earl of Orrery claimed that part 3 was "in general written against chymists, mathematicians, mechanics, and projectors of all kinds." See John Boyle, *Remarks on the Life and Writings of Dr. Jonathan Swift*, ed. Joao Froes (Newark: University of Delaware Press, 2000), 184.

2. Marjorie Nicolson and Nore M. Mohler, "The Scientific Background of Swift's Voyage to Laputa," *Annals of Science* 2, no. 3 (1937): 301.

3. Jonathan Swift, *Gulliver's Travels*, ed. Albert J. Rivero (New York: Norton, 2002), 154.

4. Isaac Newton, "Author's Preface to the Reader," in *The Principia: Mathematical Principles of Natural Philosophy*, trans. I. Bernard Cohen and Anne Whitman, assisted by Julia Budenz (Berkeley: University of California Press, 2016), 382.

5. Mary Poovey, *A History of the Modern Fact: Problems of Knowledge in the Science of Wealth and Society* (Chicago: University of Chicago Press, 1998), 97; Steven Shapin and Simon Schaffer, *Leviathan and the Air-Pump: Hobbes, Boyle, and the Experimental Life* (Princeton, NJ: Princeton University Press, 1985), 22.

6. Tita Chico, *The Experimental Imagination: Literary Knowledge and Science in the British Enlightenment* (Stanford, CA: Stanford University Press, 2018), 11. Chico's reading is part of a general impulse to re-adopt Swift as a prophet of science. Eric A. Weiss discusses Swift in computing history in "Anecdotes: Jonathan Swift's Computing Invention," *Annals of the History of Computing* 7, no. 2 (1985): 164. John McKenny proposes Swift as an early inventor of algorithms in "Swift's Prescience: A Polite Precursor of Corpus Linguistics," *Journal of Language and Literature* 2, no. 1 (2003): 48.

7. Gregory Lynall, *Swift and Science: The Satire, Politics, and Theology of Natural Knowledge, 1690–1730* (New York: Springer, 2012), 13.

8. Other books printed for Benjamin Motte include *A New Theory of the Earth, from Its Original, to the Consummation of all Things* by William Whiston, which claims to prove that the Bible's take on creation is "perfectly agreeable to Reason and Philosophy," and another by Andrew Motte, a mathematics lecture given at Gresham College titled *A Treatise of the Mechanical Powers, Wherein the Laws of Motions, and the Properties of Those Powers Are Explain'd and Demonstrated in an Easy and Familiar Method* (London: Motte, 1733), examined at the Victoria and Albert Museum, London, July 2019.

9. Allison Muri, *The Enlightenment Cyborg: A History of Communications and Control in the Human Machine, 1660–1830* (Toronto: University of Toronto Press, 2007), 101, 118; Simon Schaffer, "Enlightened Automata," in *The Sciences in Enlightened Europe*, ed. William Clark, Jan Golinski, and Simon Schaffer (Chicago: University of Chicago Press, 1999), 126–127.

10. Jennifer L. Lieberman, "Finding a Place for Technology," *JLS: Journal of Literature and Science* 10, no. 1 (2017): 27.

11. Schaffer, "Enlightened Automata," 128. Vaucanson parlayed renown across Europe for his theatrical automata into an appointment as the inspector of France's national silk trade. In that role he designed automatic silk-weaving machinery using the rotating barrels that he first designed for his automata. Edmund Cartwright claims to have invented his power loom after hearing about a chess-playing automaton. Edward Baines, *History of the Cotton Manufacture in Great Britain* (London: H. Fisher, R. Fisher, and P. Jackson, 1835), 229.

12. Barbara Hahn, *Technology in the Industrial Revolution* (Cambridge: Cambridge University Press, 2020), 2–6; Gillian Cookson, *The Age of Machinery: Engineering the Industrial Revolution, 1770–1850* (Suffolk, UK: Boydell, 2018), 226. Cookson writes that the textile industry was "curiously slow in adopting its own technology," and only a small percentage of innovations were ever patented. See also Trevor Griffiths, Philip A. Hunt, and Patrick K. O'Brien, "Inventive Activity in the British Textile Industry, 1700–1800," *Journal of Economic History* 52, no. 4 (1992): 882; Richard J. Sullivan, "Patent Counts and Textile Invention: A Comment on Griffiths, Hunt, and O'Brien," *Journal of Economic History* 55, no. 3 (1995): 666; and Robert C. Allen, "The Industrial Revolution in Miniature: The Spinning Jenny in Britain, France, and India," *Journal of Economic History* 69, no. 4 (2009): 903–904. Meanwhile, rumors had long circulated of automated looms in other countries. For instance, Thomas Powell, in *Humane Industry; or, A History of the Manual Arts, Deducing the Original, Progress, and Improvement of Them* (London, 1661), tells the story of an invention in Poland that weaved "4 or 5 Webs at a time without any humane help; it was an Automaton or Engine that moved of itself and would work night and day." See Marjorie Hope Nicolson and David Stuart Rodes, introduction to *The Virtuoso* by Thomas Shadwell (Lincoln: University of Nebraska Press, 1966), xxvn.

13. Joseph Loewenstein, *The Author's Due: Printing and the Prehistory of Copyright* (Chicago: University of Chicago Press, 2002), 119

14. The Statute of Anne (1710) granted authors fourteen years of "sole right and liberty of printing," a term limit later reinforced by *Donaldson v. Becket* (1774). As William Blackstone observed, these term limits "appear to have been copied from the exception in the Statute of Monopolies . . . which allows a royal patent of privilege to be granted for fourteen years to any inventor of a new manufacture." See Blackstone, *The Oxford Edition of Blackstone's Commentaries on the Laws of England: Book II: Of the Rights of Things* (Oxford: Oxford University Press, 2016), 405–407.

15. Swift, *Gulliver's Travels*, 29, 45, 22, 76.

16. *OED Online*, s.vv. "stocking-weaver," "stocking-frame" (Oxford: Oxford University Press), https://www.oed.com.

17. Jonathan Sawday, *Engines of the Imagination: Renaissance Culture and the Rise of the Machine* (London: Routledge, 2007), 4–6.

18. Hahn, *Technology in the Industrial Revolution*, 69. Cookson suggests that Lee's invention would have been "totemic" to readers as a symbol of labor's new social reorganization (*The Age of Machinery*, 8–9).

19. Cookson, *The Age of Machinery*, 5–13.

20. Hahn, *Technology in the Industrial Revolution*, 155; David T. Jenkins, *The Cambridge History of Western Textiles* (Cambridge: Cambridge University Press, 2003), 723.

21. The 1702 Derby mill was originally founded by Thomas Cotchett and driven by a waterwheel built by George Sorocold. This mill quickly failed, but it inspired Thomas and John Lombe's more successful mill at the same location. The Lombes, using improvements on silk throwing observed in Italy, opened the new mill in 1718 with Sorocold as engineer. This latter mill, which was five stories tall and employed over three hundred workers, is far more famous than the original, having been written about by Daniel Defoe in 1724 in *A Tour through the Whole Island of Great Britain*, ed. Pat Rodgers (London: Penguin, 2005), 72–73, and by John Theophilus Desaguliers, assistant to Newton and frequent contributor to the *Philosophical Transactions*, in *A Course of Experimental Philosophy*, vol. 1 (London: W. Innys, 1745), 69.

22. Swift, *Gulliver's Travels*, 141.

23. Lynall, *Swift and Science*, 10.

24. Cookson, *The Age of Machinery*, 16–19; Hahn, *Technology in the Industrial Revolution*, 58–62.

25. Letters discussed here include "A proposal for the universal use of Irish manufacture" (1720); "The present miserable state of Ireland" (1726), in which Swift describes Ireland's plight to Sir Robert Walpole; and "A Letter to the Archbishop of Dublin, concerning the Weavers" (1729). These letters were examined at the Victoria and Albert Museum, London, July 2019.

26. Swift, *Gulliver's Travels*, 66.

27. Lynall, *Swift and Science*, 110. The "key" referenced is Corolini di Marco's *The Flying Island, &c. Being a Key to Gulliver's Voyage to Laputa, Balnibarbi, Glubbdubdrib, Luggnag, and Japan* (London: Edmund Curll, 1726).

28. Swift, *Gulliver's Travels*, 153.

29. Swift, *Gulliver's Travels*, 153.

30. Ellen Harlizius-Kluck, "Weaving as Binary Art and the Algebra of Patterns," *TEXTILE* 15, no. 2 (2017): 180. The technical descriptions of these looms is recorded inconsistently across eras. Harlizius-Kluck highlights moments when histories disagree.

31. Schaffer, "Enlightened Automata," 144.

32. Swift, *Gulliver's Travels*, 52, 86, 112.

33. Ibid., 114.

34. Ibid., 154–155.

35. Jonathan Swift, *The Works of the Rev. Jonathan Swift, DD . . . with Notes, Historical and Critical*, ed. Thomas Sheridan and John Nichols, vol. 3 (London: J. Johnson, 1808), 267.

36. Joseph Drury, *Novel Machines: Technology and Narrative Form in Enlightenment Britain* (Oxford: Oxford University Press, 2017), 27, 85.

37. Matthew L. Jones, *Reckoning with Matter: Calculating Machines, Innovation, and Thinking about Thinking from Pascal to Babbage* (Chicago: University of Chicago Press, 2016), 59, 142.

38. Pamela McCorduck, *Machines Who Think: A Personal Inquiry into the History and Prospects of Artificial Intelligence* (Natick, MA: A. K. Peters, 1979), 317n11, 37 (Rabelais and Bacon quotes). McCorduck points to Llull rings but also the North African *zairja*, Chinese *I-Ching*, and other inventions by Pope Sylvester II, Leone Battista Alberti, and Georg Philipp Harsdorfer (9–10, 37).

39. Swift, *Gulliver's Travels*, 154.

40. Ibid., 158.

41. Jonathan Swift, *A Tale of a Tub, to Which Is Added The Battle of the Books and the Mechanical Operation of the Spirit*, 2nd ed., ed. A. C. Guthkelch and D. Nichol Smith (Oxford: Clarendon Press, 1958), 125–126.

42. Swift, *A Tale of a Tub*, 234.

43. Swift, *Gulliver's Travels*, 155.

44. See Johannah Rodgers, "The Genealogy of an Image, or, What Does Literature (Not) Have to Do with the History of Computing? Tracing the Sources and Reception of Gulliver's 'Knowledge Engine,'" *Humanities* 6, no. 85 (2017): 2.

45. John Hughes, *The Spectator*, no. 220 (November 12, 1711), ed. Richard Steele, in *The Spectator*, vol. 3., ed. George A. Aitken (New York: Longmans, Green, and Co., 1898), 246–248. Hughes also relates Peter's project to John Napier's *Logarithmorum Canonis Descriptio* (1614).

46. Matthew Jones writes that the classical rhetorical tradition, still prevalent in the eighteenth century, predicated creativity on "the putting together of previously existing ideas in novel, but not entirely original, forms" (*Reckoning with Matter*, 227–229). Lynall points out, however, that Swift's attacks on the printing industry often describe how "hack authors infus[e] their own crude productions with plagiarized material" (*Swift and Science*, 6–7).

47. Alexander Pope, *The Works of Alexander Pope, Esq., in Verse and Prose: Containing the Principal Notes of Drs. Warburton and Warton*, ed. William Lisle Bowles, William Warburton, and Joseph Warton, vol. 6 (London: J. Johnson, 1806), 251–253.

48. Brad Pasanek and Chad Wellmon, "The Enlightenment Index," *The Eighteenth Century* 56, no. 3 (2015): 361. See also Ann Blair and Peter Stallybrass, "Mediating Information, 1450–1800," in *This Is Enlightenment*, ed. Clifford Siskin and William B. Warner (Chicago: University of Chicago Press, 2010), 140; Markus Krajewski, *Paper Machines: About Cards & Catalogs, 1548–1929*, trans. Peter Krapp (Cambridge, MA: MIT Press, 2011), 54.

49. Pasanek and Wellmon, "Enlightenment Index." Samuel Johnson used slips to compose *A Dictionary of the English Language* (1755). William Smellie cut and pasted from existing works to compile the first edition of the *Encyclopedia Britannica* (1768–1771).

50. Kelly Swartz, "The Maxims of Swift's Psychological Fiction," *Eighteenth-Century Fiction* 30, no. 1 (2017): 12.

51. Swift, *A Tale of a Tub*, 96.

52. Krajewski, *Paper Machines*, 53–57.

53. The diagram of Swift's engine was first used in the second edition of *Gulliver's Travels*, published by Benjamin Motte in 1726. There is no attribution.

54. Swift, *Gulliver's Travels*, 154.

55. The letter, sent in 1736, is quoted in Paddy Bullard and James McLaverty, introduction to *Jonathan Swift and the Eighteenth-Century Book*, ed. Paddy Bullard and James McLaverty (Cambridge: Cambridge University Press, 2013), 19.

56. Hughes, *The Spectator*, 248.

57. Swift, *Gulliver's Travels*, 154–155.

58. Ibid., 52–53.

59. Drury, *Novel Machines*, 25–26. The works mentioned were published between 1760 and 1780.

60. Muri, *The Enlightenment Cyborg*, 148.

61. Swift, *A Tale of a Tub*, 269, 101.

62. Swift, *Gulliver's Travels*, 155.

63. Mark Rose, *Authors and Owners: The Invention of Copyright* (Cambridge, MA: Harvard University Press, 1993), 34, 72–74. The quote is from Baron James Eyre, in *Cases of the Appellants and Respondents*, regarding *Donaldson v. Becket* (1774).

64. Karl Marx famously flips this construction on its head in *Capital* when he calls John Milton an "unproductive worker" in the composition of *Paradise Lost*, whereas "a writer who turns out work for his publisher in factory style is a productive worker." Marx, *Capital*, vol. 1, trans. Ben Fowkes (London: Penguin, 1990), 768n, quoted in Schaffer, "Enlightened Automata," 131.

65. Jones, *Reckoning with Matter*, 99, 8, 34.

66. Joanna Stalnaker, *Unfinished Enlightenment: Description in the Age of the Encyclopedia* (Ithaca, NY: Cornell University Press, 2010), 15. See also Schaffer, "Enlightened Automata," 126.

67. Baines, *History of the Cotton Manufacture*, 229.

68. Allen, "The Industrial Revolution in Miniature," 920.

69. Desaguliers, *A Course of Experimental Philosophy*, 69.

70. Swift, *Gulliver's Travels*, 137.

71. Jones, *Reckoning with Matter*, 17, 217.

72. Stalnaker, *Unfinished Enlightenment*, 110n26. See also Schaffer, "Enlightened Automata," 129.

73. Rodgers, "The Genealogy of an Image," 4–6.

74. Hughes, *The Spectator*, 248.

75. Pope, *The Works of Alexander Pope*, 260.

76. Swift, *Gulliver's Travels*, 154.

77. Swift, *A Tale of a Tub*, 269.

78. Adam Ferguson, *An Essay on the History of Civil Society, 1767* (Piscataway, NJ: Transaction Publishers, 1980), 182–183; Schaffer, "Enlightened Automata," 129.

79. Muri, *The Enlightenment Cyborg*, 142.

80. Allen, "The Industrial Revolution in Miniature," 904.

81. Samuel Johnson, *A Dictionary of the English Language: In Which the Words Are Deduced from Their Originals, Explained in Their Different Meanings, and Authorized by the Names of Writers in Whose Works They are Found*, ed. Alexander Chalmers (London: C. and J. Rivington, 1824), 329.

82. R. S. Crane, "The Houyhnhnms, the Yahoos, and the History of Ideas," in *Reason and the Imagination: Studies in the History of Ideas, 1600–1800*, ed. J. A. Mazzeo (New York: Columbia University Press, 1962), 232.

83. Muri, *The Enlightenment Cyborg*, 142.

84. Alexander G. Weheliye, *Habeas Viscus* (Durham, NC: Duke University Press, 2014), 5–9. Weheliye, building on Sylvia Winter's and Hortense Spillers's pretense that science is often a social text, argues that there are many different, constructed definitions of "the human," as well as many discursive super- and subhumans, generated by one's positionality in regard to animality, machines, or racial norms. See also Zakiyyah Iman Jackson, *Becoming Human: Matter and Meaning in an Antiblack World* (New York: New York University Press, 2020), 23–24. Jackson adds that the liberal human was not necessarily defined against the animal but against non-European forms of rational humanism and that the figure of the animal was used to not to dehumanize but to *bestially humanize* certain humans in order to cast them as rational in the wrong way. In this essay I similarly treat the category of human thinking as moldable by the people in power.

85. N. Katherine Hayles, *How We Became Posthuman: Virtual Bodies in Cybernetics, Literature, and Informatics* (Chicago: University of Chicago Press, 1999), 3.

86. Julien Offray de La Mettrie, *La Mettrie: Machine Man and Other Writings*, ed. Ann Thomson (Cambridge: Cambridge University Press, 1996), 13.

CHAPTER 5

Designing the Enlightenment Anthropocene

Kevin MacDonnell

On April 28, 1784, James Watt acquired a patent for design improvements to the steam engine.[1] This was not the first patent that Watt and his partner, Matthew Boulton, had been issued for their work on the steam engine, nor would it be the last. It would, however, be the most impactful. Earlier steam engines had been used primarily to pump water out of mines, but the series of improvements that culminated in the 1784 patent converted the machine into a power source that could impel other machinery, a development that would position the engine as the prime mover of Britain's emerging industrial apparatus. Dating back to Karl Marx's claim that industrial capitalist production originates in "the patent that [Watt] took out in April 1784," the historical significance of this document has not been understated.[2] Marx's designation of the patent as the starting point of industrial capitalism has more recently been adopted by scientists and humanists attempting to pinpoint the onset of the Anthropocene, a term coined by the atmospheric chemist Paul Crutzen and the ecologist Eugene Stoermer to denote the current geological epoch in which humanity acts as a geophysical force.[3] And following Crutzen and Stoermer's assertion that the Anthropocene *begins* with the invention of Watt's 1784 model, the environmental philosopher Timothy Morton went a step further, saying, "We can be uncannily precise about the date on which the world ended. . . It was April 1784, when James Watt patented the steam engine, an act that commenced the depositing of carbon into Earth's crust."[4] Proponents of an eighteenth-century Anthropocene, then, implicitly contend that it is the diagrammatic expression of the steam engine's design—and not the onset of large-scale fossil fuel consumption—that triggers a fundamental shift in the relationship between humans and the natural world.[5]

That such epochal change has been attributed to the publication of a patent rather than, say, the installation of these technologies suggestively foregrounds the role of design in producing the material conditions of the Anthropocene. Patents are, first and foremost, legal documents. But they can also mark the arrival of an idea or concept as a discursive actor, often representing their first articulation in the public sphere. This is certainly the case with Watt's improvements to the steam engine. Although it would be decades before the machine became ubiquitous in British manufacturing, its key design features were codified in the 1784 patent. As such, the very notion of an eighteenth-century Anthropocene places eighteenth-century intellectual culture firmly at the center of modern environmental history. Or as Alan Mikhail argues in his essay "Enlightenment Anthropocene," the contemporaneity of an eighteenth-century Anthropocene and a major intellectual movement like the Enlightenment is a "coincidence" that necessitates greater investment in efforts to understand the latter's role in producing the former.[6]

This chapter seeks to answer Mikhail's call. In what follows I unpack the sociocultural contexts out of which Watt designed the 1784 patent, which can in turn amplify the conceptual foundations subtending industrial capitalism and the Anthropocene. The key features introduced in the patent, I argue, extend on the artisanal theory of design spearheaded by the publication of William Hogarth's aesthetic treatise *The Analysis of Beauty* (1753). I focus my analysis on how Hogarthian aesthetic and epistemological commitments informed the most important addition to the 1784 patent, an innovative mechanism known as "Watt's linkage," the introduction of which established the steam engine as what Marx calls "an agent universally applicable in industry" (figure 5.1).[7]

Designed to synthesize the linear motion of the engine's piston rod and the curvilinear motion of its working beam, Watt's linkage allowed the steam engine to produce a steady supply of mechanical force by translating these competing trajectories into what would become the machine's signature reciprocating flow. The linkage achieved this synthesis by tracing an undulating line that one nineteenth-century historian of the steam engine characterized as "a species of S-curve."[8] I contend that the distinctive S-curve generated by Watt's linkage is born out of the same epistemological foundations that underpin the serpentine "line of beauty," the aesthetic principle at the heart of Hogarth's philosophy of design. While the morphological affinity between the two is striking, their superficial resemblance means little in itself. The serpentine figures expressed by Watt's linkage and Hogarth's line of beauty are, more importantly, representative manifestations of a shared philosophy of design that constituted aesthetic and technological production amid Britain's industrial transition.

During the second half of the eighteenth century, the serpentine line emerged as a potent symbol of artisanal efforts to navigate the limitations of classical geometrical design. Euclidean geometry had long been employed as the dominant conceptual framework for both the fine arts and mechanical arts, but its dependence on metaphysical abstractions as the grounds for design came under assault as British epistemology became increasingly committed to empirical analysis throughout the period.[9] The precipitous rise of empiricism brought with it a set of design practices that privileged material application over abstract formulation. If the straight line and the circle reflect the ideal figures of classical geometry, then the serpentine line embodies something wholly different, capturing the organic

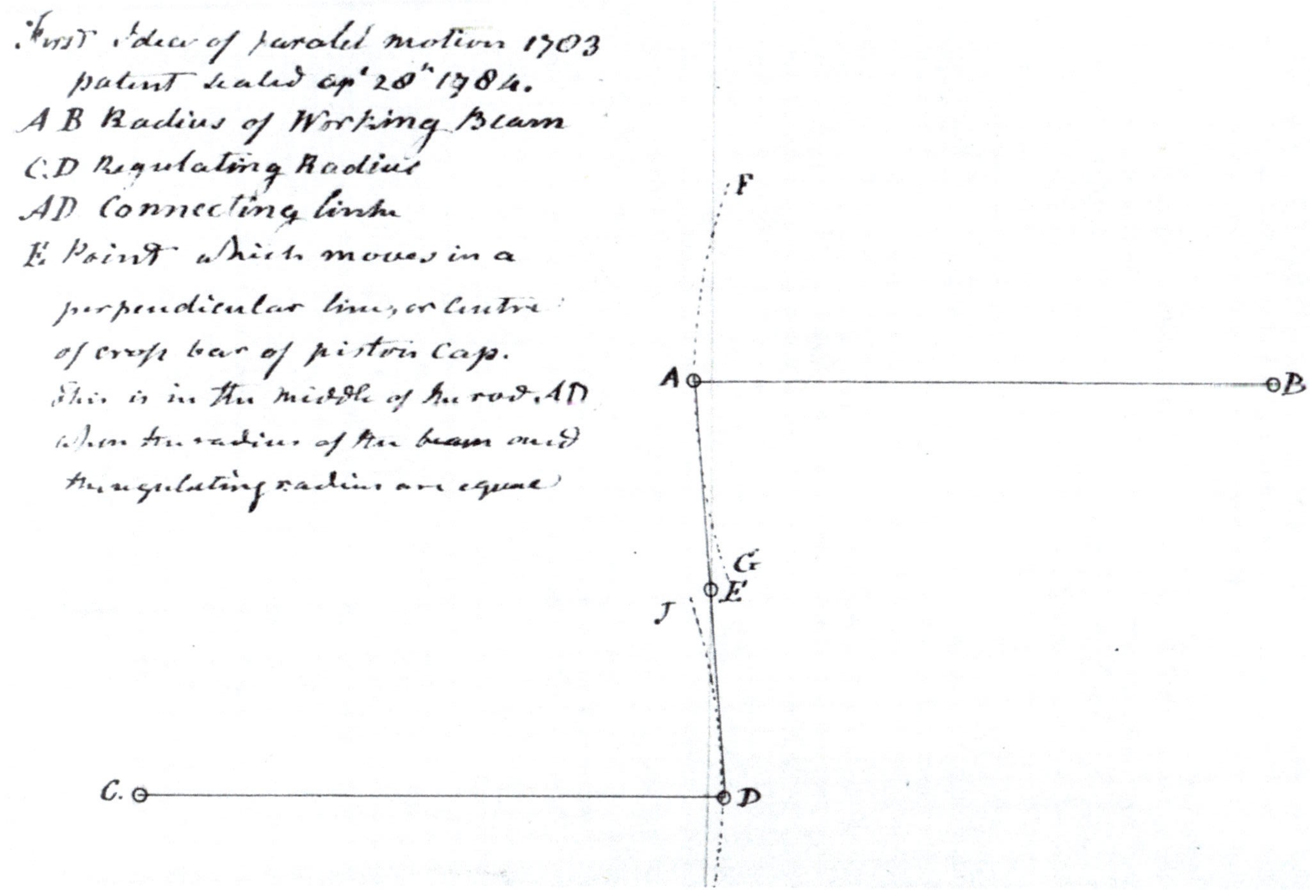

Figure 5.1. Sketch of Watt's linkage, James Watt to Matthew Boulton (1784). Reprinted from *The Selected Papers of Boulton and Watt*, vol. 1, *The Engine Partnership, 1775–1825* (Cambridge, MA: MIT Press, 1981).

fluidity that constitutes material form. This turn toward more empirical design practices that is reflected in the irregular form of the S-curve proved critical to the success of industrialization efforts. By attending to material rather than metaphysical notions of formal design, engineers like Watt and artists like Hogarth were able to effectively harness a natural world that exceeded linear calculation.

Given the position of the steam engine within the historiography of the Anthropocene, my discussion of the shared principles of design found in Watt's linkage and Hogarth's line of beauty inevitably complicates the legacies of the latter. If it is indeed the diagrammatic expression of these design principles that marks the onset of the Anthropocene, then it follows that their manifestation in contemporary aesthetic forms played a part in the sociocultural transformation that accompanied this new geological epoch. However, lest we too quickly indict eighteenth-century aesthetic philosophy for aiding and abetting the onset of the Anthropocene, I conclude this chapter by considering the application of these design principles in ways that undermine the instrumental logic of the industrial era. To do so, I examine Laurence Sterne's novel *Tristram Shandy* (1759–1767), a text that identifies itself as a "machine" and does not shy away from its indebtedness to Hogarthian aesthetics and whose serpentine narrative offers an alternative iteration of the innovation in question. By centering my inquiry around a set of common practices applied within engineering, art, and literature, I therefore track the role of Enlightenment-era design not only in shaping and expressing the Anthropocene's dominant modes of knowing and making but also in challenging their implementation.

The serpentine style had been incorporated into artistic practice dating back to the Renaissance-era *figura serpentinata*, but it was during the middle decades of the eighteenth century that the undulating S-curve rose to prominence within Britain's arts establishment, assuming a central position in the artisanal design practices that opposed the abstract conventions of high art.[10] Up until midcentury, architectural and artistic design had largely conformed to the neoclassical theory formalized by Francis Hutcheson and Lord Shaftesbury, who equated "the Beautiful" with the ideal figures of classical geometry.[11] The formal integrity that Hutcheson ascribes to hexagons, octagons, or decagons, for instance, is derived from their correspondence to conceptual figurations explicitly divorced from material reality. Straight lines and perfect circles do not, after all, appear in nature. This unnatural quality of geometrical form stood out as a testament to the brilliance of human artifice, with the straight line appearing as what Tim Ingold describes as "an index of the triumph of rational, purposeful design over the vicissitudes of the natural world."[12] Significantly, the basis for such thinking was decidedly abstract, centralizing metaphysical form as the grounds for aesthetic design.

The subsequent resurgence of the serpentine style occurred in response to this glowing appraisal of classical geometrical form within high art. W.J.T. Mitchell concedes that the S-curve is itself a geometrical figure—a member of the vortex family—though he is quick to note that its divergence from "pure" geometrical forms like the straight line and circle in late eighteenth-century aesthetic and mathematical theory is "emblematic of the historic shift from classic to modern geometry."[13] Whereas Britain's early eighteenth-century aestheticians endorsed the regularity of straight lines and circles, those who worked in more artisanal fields like furniture and landscape design began to "show," in William Shenstone's words, "to the pupils of Design, / The triumphs of the waving line."[14] The figure initially appeared as a central component of the rococo style imported from France around midcentury, visible in the Rocaille's utilization of serpentine lines to portray what Siegfried Giedion calls the "meeting of reason with the richness of organic form."[15] The serpentine style ultimately achieved influence within British aesthetics in landscape and garden design, the practitioners of which sought to imitate the flexibility of natural form as a counterbalance to the stifling formality of neoclassicism. This contrast is evident in the work of William Kent, whose approach to landscape design stems from the position that "Nature abhors a straight line," even as his contributions to architecture adopt the linear Palladian style.[16] Likewise, the asymmetrical S-curve was notably a preferred feature in the elaborate gardens designed by Capability Brown, whose use of the serpentine style was widely adopted by his contemporaries and successors throughout the eighteenth century.[17]

After initially gaining traction in the domain of landscape architecture, the serpentine line would receive its most elaborate theorization as a design principle in Hogarth's *The Analysis of Beauty*, in which the waving "line of beauty" is declared to be the cornerstone feature of aesthetic value (figure 5.2). Conceived by Hogarth as a challenge to the precision of academic design theory, he crafts a system of design around the line of beauty, to which, as he has it, "all mathematical schemes are foreign."[18] A fundamentally empirical treatise, the *Analysis* rejects the geometrical aesthetics of Shaftesbury and Hutcheson on the basis of Hogarth's observation of the striking dissonance between the mathematical calculability of neoclassical design and the "variety" and "intricacy" of form that has a "real foundation in nature" (33).[19] Hogarth, who posits the human body as the quintessential figure of "native beauty," recognizes that the body does not align with neoclassical conventions: "no exact mathematical measurement by lines can be given for the true proportions of a human body" (39). Moreover, attempts to contort the body to "mimic such very straight or round motions" would be "incompatible with the human form" and "therefore ridiculous" (106). Individual body parts like bones and organs might be rigid or visually unappealing, but Hogarth contends that our "elastic skin" connects these components into a smooth, curving body (56). Michael Baridon has argued that the scientific associations with Hogarth's use of "elastic" here inherently challenge the "omnipotence" of geometry, "for the expansible character of gases made it impossible to represent them by geometric shapes."[20] From gaseous substances to human bodies, empirical inquiry was revealing the surrounding world to be fluid and amorphous, leading Hogarth to suspect that contemporary design was hampered by a fundamental misapprehension of reality. Or as Baridon says, "Geometrical figures were too rigid to express the flexibility of living forms."[21] Hogarth's argument in the *Analysis* follows from this insight, as he goes on to articulate a system of design that is attuned to the elasticity of nature rather than the "mean lines" of abstract forms (106).

THE

ANALYSIS

OF

BEAUTY.

Written with a view of fixing the fluctuating IDEAS of TASTE.

BY *WILLIAM HOGARTH.*

So vary'd he, and of his tortuous train
Curl'd many a wanton wreath, in sight of Eve,
To lure her eye.-------- Milton.

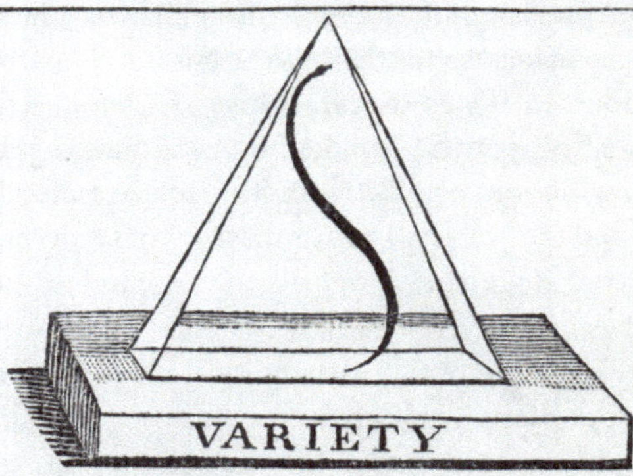

LONDON:

Printed by *J. REEVES* for the *AUTHOR,*
And Sold by him at his House in LEICESTER-FIELDS.

MDCCLIII.

Figure 5.2. Frontispiece from William Hogarth's *The Analysis of Beauty* (London, 1753). Courtesy of the University of Wisconsin Digital Collections Library.

Although Hogarth's critique is primarily directed at the fine arts, his diagnosis of the misguided commitment to such abstractions extends to his views on the mechanical and industrial arts as well. Hogarth's perspective on technological design can be gleaned from the work of one of his disciples, Jean Andre Rouquet, whose *The Present State of the Arts in England* (1755) has been treated as a recapitulation of the artist's own views.[22] Based on his examination of the production methods used within skilled labor enterprises ranging from painting and architecture to silk manufacture and steel work, Rouquet's treatise concludes that the English, while valuing design, were hampered by a "geometrical rigour" that restricts the potential of material production.[23] This approach, Rouquet argues, "must surely contribute to the improvement of our reason: but everything may be abused; we may reason too much; thence may be too vain, and come to despise everything that does not appear exactly reasonable."[24] Machines and instruments built according to strict mathematical laws thus reflect, for both Rouquet and Hogarth, a fundamental incommensurability with the material world. In the *Analysis*, clockwork mechanisms like John Harrison's chronometer and Jacques Vaucanson's mechanical duck are characterized as "confused" and "displeasingly shaped" because of the unnatural movement of their rigid components (62). These machines, like the geometrical figures endorsed by Hutcheson, achieve their desired ends yet fail to consolidate their internal motion into the fluid regularity found in nature. Hogarth's vocal critique of contemporary technology notwithstanding, he does not go so far as to claim that the arts could not construct a figure with both graceful motion and practical application—"variety" *and* "fitness"—only that within Britain's contemporary regime of material production, they largely had not. Hogarth's primary aim in the *Analysis*, then, is to cultivate the design principles that could be employed to build such "machines."

In response to Hogarth's dissatisfaction with the geometrical straitjacket of contemporary design, he conceives the line of beauty as a technique that posits "variety" of motion as the founding principle of material production. The basis of his valuation of the line of beauty lay in its application outside the strictures of geometrical form, appearing instead as a founding example of what Scott MacKenzie calls the "post-classicist line" that was "no longer determined and explicable, *a priori* by geometrical principle."[25] Whereas motion had long been taken as a direct product of linear composition, Hogarth flips the script and identifies linear form as a by-product of motion. On the basis of this assumption, he offers the continuous undulating motion of the line of beauty as the central component of aesthetic beauty. By forcing the viewer's eye to physically mimic the reciprocating trajectory of an S-curve, the line of beauty appeals to the mind's "love of pursuit" by "*lead*[ing] *the eye a wanton kind of chace*, and from the pleasure that gives the mind, intitles it to the name of beautiful" (33). As such, the line of beauty challenges the systematicity of geometrical design by mobilizing the indeterminacy of empirical reality, representing an aesthetics that is, as Timothy Erwin describes it, "obstinately of this world."[26]

Hogarth's attention to variety, however, does not come at the cost of legibility. Conceding that "variety uncomposed, and without design, is confusion and deformity," he advocates for a "composed variety" that is embodied in the line of beauty's irregular S-curve (28). Proper design requires a degree of uniformity to accompany variety, so rather than letting variety run amok, "Simplicity is call'd in to restrain its superfluities" (39). The line of beauty, a figure composed of any "two curves contrasted," appears at the intersection of classical and modern geometrical form, representing for Hogarth the mediation of the regularity of artifice and the elasticity of experience (42). This mediating quality is invoked by Ronald Paulson, who describes the artist's approach to design as a "middle way," sitting firmly "between the static Beautiful of Shaftesbury and Hutcheson and the uncontrollable chaos of the Sublime."[27] Echoing Paulson, Mitchell refers to Hogarth's "balancing of antithetical values," while Frédéric Ogée describes the line of beauty as "halfway between abstinence and excess, between the rigid straight line of dogmatism and the excessively curved line of hedonism."[28] This strategic retention of discipline to balance—or, as Paulson says, "regulate"—the flexible nature of material reality is where Hogarth most radically diverges from his predecessors and what gives the line of beauty its distinctive character.[29]

Despite Hogarth's own conflicting attitudes toward the mechanical arts, the design principles he establishes in the *Analysis* were received warmly among craftsmen.[30] In fact, the relegation of the *Analysis* to mere "art theory" is more a historiographical misconception based on Hogarth's success as a painter than an accurate appraisal of the text. That Hogarth saw the wider application of his design theory is evident in his decision to send a copy of the *Analysis* to the Royal Society, Britain's premier scientific institution, along with his rejection of the line of beauty's association with objects that are "no way applied, nor of any manner of use, but merely to entertain the eye" (60). Unlike the serpentine lines used within rococo aesthetics, Hogarth's line of beauty "covered much more territory," emerging as a general-purpose principle of material production that could be adopted by mechanics as well as painters.[31] Indeed, long after Romantic treatises on aesthetics had supplanted Hogarth's work, the line of beauty remained a useful design principle within nineteenth-century engineering. An 1833 issue of *Mechanics Magazine*, for instance, invokes the line of beauty as evidence of the superior design of the latest stagecoach model. Unlike older models, the axle in these new stagecoaches implements a "return curve (Hogarth's line of beauty)," which "is a tolerable evidence that the builders are not totally ignorant of the rules by which beauty is produced."[32] Likewise, in an account of the *Brougham* and *Lowther* steam locomotives, the British railway writer C. Hamilton Ellis describes the "sweep" of the engine's cylinder as "that *Line of Beauty and Grace* which Hogarth drew on the palette of his self-portrait in 1745."[33] The line of beauty's versatility in this respect can be attributed to what Ruth Mack identifies as Hogarth's "practical aesthetics." Having "straddled the role of artist and artisan for his entire career," Hogarth derives his theory of design from "the artisanal domain of habitual action," grounding his notions of form, function, and beauty in practice.[34] Although the scholarly conversation regarding the line of beauty's application has largely been restricted to its use as a principle of artistic production, reading Hogarth artisanally, as Mack does, allows us to reorient the line of beauty in relation to contemporary developments in technological design.

Considered as a feature of technological as well as aesthetic design, the line of beauty's manipulation of empirical reality reflects the governing logic of modern technology, as defined by Martin Heidegger. The "revealing" of essence that "holds sway" in modern technology occurs through what Heidegger calls a "challenging," which means "to call forth" or "to demand positively."[35] Whereas a Dutch windmill's ability to perform work is passively "left entirely to the wind's blowing," a hydroelectric plant "sets upon" a river by appropriating its energetic potential and then directing it toward something else.[36] Pre-industrial machines, like Vaucanson's clockwork duck, operate irrespective of existing physical forms, manifesting themselves as material representations of strictly abstract principles. "Modern" technology, on the other hand, departs from its predecessors by actively exploiting the material world. In this sense, the line of beauty *behaves* like a feature of modern technology, manipulating the intricate world of experience—empirical reality—to construct culturally legible forms. If employed properly, or "artfully," argues Hogarth, the line of beauty has the ability to "set native beauty off to more advantage" (39). Its mediating or regulatory approach to design attends to the unwieldy variety of nature and strategically formalizes this quality rather than surrendering to it. Insofar as the line of beauty approaches empirical reality as a dynamic force that can be molded into a coherent figure, the system of design that Hogarth lays out participates in the same will to regulate or control nature that has come to define technology in the Anthropocene. Different in degree only, the extraction and assimilation that characterizes technological activity in the Anthropocene is visibly at work in Hogarth's articulation of the line of beauty. This is certainly not to say that the art created by Hogarth or inspired by the *Analysis* is any way responsible for the environmental destruction brought about by industrial capitalism amid the Anthropocene. I am, however, suggesting that the cultivation and reification of strategies and practices within the realm of the aesthetic should be accounted for in any analysis of the conceptual foundations of the Anthropocene.

It is against the backdrop of Hogarth's ranging influence on design that we can situate the steam engine that Watt patented in 1784, itself one of Heidegger's core examples of modern technology and a machine whose invention has been directly tied to the onset of industrial capitalism and the Anthropocene. Like Hogarth, Watt was successor to a mathematical system of design founded on the abstract principles of Euclidean geometry. Watt, moreover, similarly responds to the inherent constraints of this framework by adopting a serpentine motion into the mechanism at the center of the 1784 patent, a technical feature that reflects his own empirical approach to design. Whether Watt was directly familiar with Hogarth's work is not only unclear but ultimately outside the scope of this study. It is rather my aim to identify their shared approaches to design as "open techniques," a term that Liliane

Hilaire-Perez and other historians of technology have used to describe the skilled practices that both technical and nontechnical actors develop and circulate throughout society. “The techniques of production,” Hilaire-Perez and Catherine Verna point out, “were not only of interest to producers, but also to merchants, shopkeepers, artists, consumers, local authorities, princes, political writers, and others. Thus the actors involved in technical dissemination should not be limited to the technicians.”[37] Visual culture emerged as an important site for such interchange, offering techniques and technologies that could be used by artists and engineers alike. Pamela Smith tracks this interdisciplinary application of visual culture back to the use of images in early modern science, arguing that artists and artisans “helped constitute the aims and methods” of empirical scientific inquiry.[38] Nor is Smith alone in this regard. Eileen Reeves and Dániel Margócsy, among others, have offered similar assessments of the relationship between visual culture and technoscience, presenting compelling evidence of the reciprocal influence that these knowledge systems have had on each other.[39] Following the emergence of the public sphere but before the erection of firm disciplinary boundaries, eighteenth-century Britain’s knowledge culture was well suited to foster such conceptual transmission. The serpentine motion that appeared as an open technique at this time can thus be positioned as a shared strategy for responding to similar epistemological challenges faced across the fine arts and mechanical arts.

In the decades that followed the invention of the steam engine, it shuttled into popular consciousness as an object whose allure extended beyond technicians and manufacturers. The dissemination of the machine initiated what Ben Russell calls a “new machine aesthetic,” in which the steam engine came to exert “a continued imaginative effect on those who saw it work.”[40] Within modern scholarship, the aesthetic quality of the steam engine has typically been discussed in these terms, as a phenomenon that proceeds from, rather than informing, the machine’s design.[41] The aestheticized perceptions of the steam engine, according to such accounts, merely extend from its function, engendered by the recognition of how serviceable the machine could be to capitalist production. But while these analyses deliver important insights into how nineteenth-century aesthetics was shaped by the invention of the steam engine, they do not consider how the steam engine may have been shaped by eighteenth-century aesthetics. As Russell reminds us, “The engine was as much a cultural machine as a scientific one,” and it is therefore imperative that we remain attentive to the discursive formations embedded within its design.[42]

The groundbreaking design improvement at the center of the 1784 patent appeared in the form of a diagram for the unassuming mechanical contraption that became known as “Watt’s linkage” (figure 5.3).[43] The unimpeded reciprocating motion generated by the linkage could be used to power any machine driven by a wheel, providing a source of mechanical force more reliable and consistent than animals, wind, or water. Two years prior, Watt had taken out a patent for the sun-and-planet gears—a device attached to the flywheel that drove the piston both up and down, making the engine “double-acting”—which had begun the process of bringing the engine out of the mine and into the factory. But without the addition of

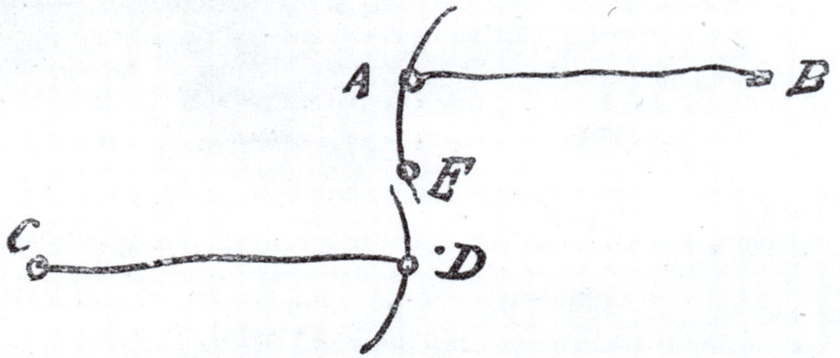

Figure 5.3. Sketch of Watt's linkage, James Watt to his son, James Watt Jr. (1808). Reprinted from Franz Reaulaux, *The Kinematics of Machinery: Outlines of a Theory of Machines*, ed. and trans. Alex B. W. Kennedy (London: Macmillan, 1876).

the linkage introduced in 1784, the engine's signature reciprocating motion would remain imperfect, severely delimiting the machine's maximum output and prohibiting its use as a prime mover.[44] Watt's linkage worked by incorporating a flexible joint that could accommodate both the angular motion of the working beam and the perpendicular motion of the piston. The joint was composed by a three-bar linkage that connected the two contrasting motions and converted them into a unified reciprocating motion. According to Helmut Müller-Sievers, synthesizing "translation" (straight-line motion) and "rotation" (circular motion) into a singular motion had long been considered a contradictory proposition within Western metaphysics. Watt's linkage, however, "forced a compromise between rotation and translation."[45] The mechanism achieved this compromise by attaching parallel bars to both the piston rod and working beam that would each trace a curvilinear line and then connecting these with a third bar. This third bar converted the two semicircles into an undulating line that produced the linkage's signature S-curve.

The serpentine line traced by Watt's linkage emerged as the central feature of the machine's aforementioned aesthetic appeal. In an address to the French Académie des sciences, François Arago's technical description of the steam engine devolves into aesthetic admiration when he arrives at the linkage: "At each reciprocation of the stroke, it opens and closes, with the smoothness—I had almost said with the grace,—which charms us in the gestures of an accomplished actor."[46] While technological design is often considered antithetical to aesthetic taste, the "grace" of the linkage's motion leads Arago to position the steam engine as evidence of his stated belief that "the useful arts may often exhibit the same beauty and greatness which are displayed in the fine."[47] Boulton and Watt clearly recognized the hypnotic appeal of the engine's smooth reciprocating motion, taking advantage of this effect by showcasing the first of these models in prominent London locations where it would be visible to the public.[48] After witnessing the mechanism in motion for the first time, Watt remarked to Boulton on his "surprise at the perfection of its action": "in looking at it for the first time, I had all the pleasure of novelty which could have arisen had it been invented by another person."[49] Though Watt's suggestion here is probably a rhetorical flourish, the steam engine's application of the reciprocating flow of the S-curve to power Britain's industrial transition is nevertheless notable given its well-established position within eighteenth-century aesthetic discourse.

In addition to the morphological affinity between Hogarth's line of beauty and Watt's linkage, Watt arrives at the serpentine line by way of the same challenge to contemporary design practices that had catalyzed its appearance elsewhere. His linkage would revolutionize manufacturing in the coming decades, but more immediately, it signaled a departure from the "geometrical rigour" that Rouquet had diagnosed within the arts. As Michel Serres notes in his account of the major conceptual paradigms before and after the Industrial Revolution, Euclidean geometry influenced the direction of pre-industrial technological design as much as it had informed pre-industrial aesthetic philosophy. As a result, the force generated by pre-industrial machines was taken to correspond directly to "their form, their lines, their geometry."[50] The world of machines before the Industrial Revolution was, in other words, "a world that is drawn, drawable. It is a world in which chains trace motion."[51] Until the 1780s, steam-engine design had reflected this prevailing logic, producing limited force through the machine's assemblage of geometrical components and thus wasting most of the energy produced in the engine's boiler. So entrenched was this approach to mechanical design that Watt's stated intent to replace the rigid contraptions used in earlier engines with the flexible linkage he devised was initially received by Boulton as "quite incomprehensible."[52] Watt's challenge to this system came in the linkage he patented in 1784, which attended to the "elastic force" of steam into the machine's technical design.[53] Against the prevailing "mechanical" theory of heat, Watt and his mentor, Joseph Black, had devised a "chemical" theory that conceived of heat as a substance that combined with other substances. On the basis of this theory, Watt concluded that steam was an "elastic fluid" that expanded when combined with heat. Such expansion, though, would "not follow a Geometrical progression," leading Watt to opt for a flexible linkage that accommodated steam's bounding movement.[54] Though the mechanism itself was rigid, composed of three metal rods, the "bewildering variety" of its serpentine motion could not be charted "algebraically or in any other form of abstraction."[55] Its motion was consistent but exceeded linear calculation. Watt's linkage thus successfully mediated the engine's rigid mechanical principles and steam's elasticity, manifesting in the mechanism's trademark serpentine motion. While his earlier improvements had effectively contained and redirected steam power, the unbroken reciprocating motion produced by the 1784 linkage allowed steam's expansive principle to produce a level of force that exceeded the limits of the machine's linear components. As such, the design of the linkage reflects the "irreducible empiricism and pragmatism" underlying Watt's work on the steam engine, a feature that is embodied in the mechanism's winding S-curve.[56]

My aim thus far has been to contextualize the "species" of motion produced by Watt's linkage by situating it in relation to the serpentine style articulated within Enlightenment-era aesthetic discourse. The shared epistemological foundations of these artifacts of eighteenth-century design do, however, complicate the historical legacies of Hogarthian design. What, exactly, happens to our understanding of eighteenth-century aesthetics once we locate its design principles within the discursive and material formations responsible for initiating industrial capitalism and, by extension, the Anthropocene? Jesse Oak Taylor has tackled this question as it pertains to the "rise of the novel," a literary-historical phenomenon that "dovetails" with the onset of industrialization, an alignment, he argues, that is "not coincidental, but constitutive."[57] As he explains, the novel's representational conventions, which centralize the individual human subject against a static environment, are built on dangerous anthropocentric notions of command over the natural world. Accordingly, he concludes, "The novel form is at once a product and a participant in the social, historic, economic, and ecological forces responsible for bringing the Holocene to an end."[58] Having identified the genre's problematic role in this regard, Taylor contends that alternative representational forms are needed to undermine the governing logics of the Anthropocene.[59] But before I condemn the *Analysis* on similar grounds for advancing techniques that would be replicated by the engines of industrialism, it is important to briefly consider the manifestation of these principles in a slightly different context. To do so, I would like to conclude by turning here to Laurence Sterne's *Tristram Shandy*, a text whose frequent appraisal of itself as a Hogarthian "machine" represents an alternative response to the limits of geometrical design heretofore not considered.

Writing in the wake of the highly systematic approaches to language issued by David Hartley, Robert Hooke, and numerous early novelists, Sterne perceived contemporary representational frameworks to be fundamentally at odds with the limitless variety of experience.[60] To paint a full picture of a "life," as *Tristram Shandy* attempts to do, exposes the limits of such enterprises. Tristram grapples with the inherent challenges of linear representation when his attempt to recount the day of his birth, which is frequently sidetracked by a number of contextual and conversational digressions, takes over a year: "at this rate I should just live 364 times faster than I should write—It must follow, an' please your worships, that the more I write, the more I shall have to write. . . . I shall never overtake myself."[61] This is after taking two chapters to recount "what passed in going down one pair of stairs."[62] Experience is presented here and elsewhere in the text in "elastic" terms, with each moment possessing the capacity to expand into any number of backstories or digressions. "To write a linear narrative is morally impossible for Tristram," argues Joseph Drury, "because to do so requires placing abusive constraints on human nature."[63] Plotting Tristram's life within a linear narrative structure would, Sterne admits, necessitate the elimination of the text's digressions, which are "the sunshine—they are the life, the soul of reading."[64] He therefore sets out to assemble a text that incorporates as central structural components the asides, tangents, and digressions that constitute experience.

Sterne embraces the unpredictable variety of experience in this regard, but he does not abandon linearity altogether. Rather, his adoption of Hogarth's principles of design appear in the very mechanism of the narrative, which embodies the line of beauty's "composed variety." In this sense, the novel is presented not as an undifferentiated collection of digressions but as a well-oiled "machine" that mediates or "reconcile[s]" the "contrary motions" of linear narrative and elastic experience into something of a serpentine trajectory. "In a word," Shandy summarizes, "the machinery of my work . . . is digressive, and it is progressive too,—and *at the same time*."[65] The degree to which he accomplishes this mediation is illustrated in the impossibility of distinguishing between linear and nonlinear moments in the text, even, as Victor Shklovsky points out, in the scenes that seem overtly digressive: "Sterne's inset material does not play a merely peripheral role in the novel. On the contrary, every passage belongs to one of the novel's compositional lines."[66] Christina Lupton has argued that the ability to recognize the novel's digressions *as* digressions requires that Sterne "first posits his work as a book," citing the linear succession of text in the material book as a precondition for *Tristram Shandy*'s nonlinear appearance.[67] The text's representation of the variety of experience is, in other words, dependent on a certain degree of regulation and control, most evident in the simultaneously linear and nonlinear plotlines Sterne provides at the end of the fifth volume. Even as Tristram facetiously scolds himself here for failing to tell any part of the story "in a tolerable straight line," his comments throughout the novel imply that he was very much pleased with the mediation of progressive and digressive elements reflected in these irregular lines.[68]

Not only does the form of the novel reflect this mediation of linear and nonlinear elements; so, too, does the composition of the Shandy family.[69] In order for Tristram to illustrate his claim that Shandy-hall constitutes a "complex" rather than a "simple machine," he describes its inner workings as resembling "the inside of a *Dutch* silk-mill." He explains this comparison by referring to the "parallel" forms of "motion, debate, harangue, dialogue, project, or dissertation" that occur simultaneously in the house's separate quarters and remain interconnected by way of "the rule to leave the door, not absolutely shut, but somewhat a-jar." This rule is enabled by "the covert of the bad hinge, (and that possibly might be one of the many reasons why it was never mended)." Information travels back and forth through the cracked door, which was "wide enough, for all that, to carry . . . this windward trade," keeping the residents of Shandy-hall on the same page. Though the household is composed of a range of actors who are "contrasted by nature and education," the door hinge mediates their incommensurable "motions," synthesizing these perspectives into a single collective experience. Sterne uses the death of Bobby Shandy to illustrate how this "family-machine" worked, presenting the different ways in which the Shandys and their servants process the tragic news as it circulates across either side of the open door. As Walter's address to the parlor attempts to rationalize the loss by recounting a linear history of the philosophical interpretation of death, "proceeding from period to period, by metaphor and allusion," Trim provides a less structured and more emotional response to the tragedy. His oration in the kitchen is "without wit or antithesis, or point, or turn, this way or that; but leaving the images on one side, and the pictures on the other, going straight forwards as nature could lead him, to the heart." Sterne's depiction of the "complex" Shandy family-machine here implies that Bobby's death could only be conceptually approached by balancing these alternative forms of expression, requiring neither Walter's rigid logos nor Trim's disjointed pathos alone. Some combination of the two, however, mediated by the mechanical hinge between them, begins to register the immensity of such an experience.[70]

Sterne's use of the language of machinery to stage his critique of the limits of representation speaks to the common structural limitations faced by mechanical production and artistic representation alike. That Sterne packages the novel as a "machine" is often read metaphorically, but Drury usefully points out that during the eighteenth century, "the novel was understood to be a kind of technology." As a result, "critical reflection on its uses and effects was necessarily implicated in contemporary discourses about the uses and effects of the Enlightenment's other machines."[71] As such, pulling apart Sterne's ideas concerning the limitations of narrative from those about technological design writ large becomes impossible. And while Roy Caldwell has argued that Sterne's machine "belongs not to the family of modern industrial machinery, but is rather the heir of the archaic machines of the seventeenth century," the text's incorporation of the stochastic power of digression as the source of its literal and figurative motion is distinctly modern.[72] Unlike the rigid mechanisms constructed by Sterne's contemporaries in the domains of literary and technological production, *Tristram Shandy* mediates linear progression and nonlinear digression to generate his machine's motive force. The middle way that the text inhabits aligns Sterne's project with those techniques employed by Hogarth and Watt insofar as it attempts to practically negotiate the systematizing nature of design with the elastic quality of empirical reality.[73]

The novel, however, puts these mediating techniques to very different ends, conceiving of itself as a machine that, while certainly functional, does not actually do or create anything. The novel is rather, as Yorick announces in the text's closing line, but "a COCK and a BULL."[74] Take the aforementioned door hinge, which importantly performs its mediating role by *not* working properly. Though classified as a "bad hinge," its dysfunction as a simple machine proves to be critical to its function as part of Sterne's complex machine. The hinge, in other words, acquires its value outside of the parameters of conventional technological design, operating irrespective of the rules established within that system. Moreover, the family-machine made possible by this broken machine does not itself function in a way that could be deemed "productive" or "progressive." The circulation of the news of Bobby's passing between the parlor and the kitchen, while allowing the characters on either side of the cracked door to begin processing the death in their own ways, nevertheless culminates in a miscommunication between Walter and Mrs. Shandy. Standing outside the door, Mrs. Shandy mistakes Walter's recitation of Socrates's statement "I have friends—I have relations,—I have three desolate children" as an admission of infidelity, prompting her to charge into the parlor to confront her husband.

—Then, cried my mother, opening the door,—you have one more, Mr. *Shandy*, than I know of.

By heaven! I have one less,—said my father, getting up and walking out of the room.[75]

Rather than facilitating clear communication between the separate rooms in the house, the door hinge here obstructs communication. By concluding the account of Bobby's death with such a farcical denouement, Sterne reveals the "complex" machine that is Shandy-hall to be inherently unproductive. Though capable of mediating the characters' contrary "motions" into a generative collectivity—Mrs. Shandy is eventually informed of her son's death—in the end, the family-machine only generates a comical moment of misunderstanding.

Against the instrumental turn that defined technological development within Enlightenment Britain, Sterne imagines the novel to be a perfect machine that cannot be incorporated into contemporary systems of production.[76] His text-machine thus comes to function as an antimachine, effectively mobilizing the elastic quality of empirical reality, but to no end that would be deemed "productive." *Tristram Shandy* simultaneously replicates and disavows the conceptual schema it had adopted from contemporary design, offering a compelling instance of Enlightenment self-critique. Consider for a moment the account of Trim's failed attempts to tell Toby "The story of the king of Bohemia and his seven castles." Every time Trim begins, Toby interjects with an irrelevant question or comment, resulting in a temporary digression away from the story. Trim begins the story from scratch five times, each beginning represented by an interruption in the text itself: "The story of the king of Bohemia and his seven castles, continued."[77] With every new attempt, Toby finds a way to derail what is presumably Trim's conventionally linear story. Sterne includes scenes like this as a way to maintain his formal balance between digression and progression, reinforcing their mediation in the text by repeatedly proceeding with the story, only to return each time to the beginning. There is no product or payoff produced by Sterne's machine here; rather, the value lies simply in witnessing its motion.

The de-instrumentalization of Hogarth's principles achieved in Sterne's text-machine suggests that such techniques are not inherently exploitative, nor must they necessarily be abandoned to challenge the conceptual underpinnings of industrial capitalism and the Anthropocene. The decline of classical geometry within aesthetics and engineering ushered in a way of knowing and making that replaced abstractions with empirical reality as its animating subject, yet the emanations of this new epistemological paradigm differed greatly across the fine arts and industrial arts, if not in theory than in application. According to Gilbert Simondon, the often-nefarious reorientation of technical practices is a common, if not constitutive, occurrence in the history of technology. "There appears to be a singular law of transformation of human thought," he argues, "according to which any ethical, technical, and scientific invention, which sets out as a means of liberation and rediscovery of man, becomes through its historical evolution an instrument that turns against its liberation and enslaves man by limiting him."[78] Eighteenth-century design theory did not set out to develop skilled practices and techniques that could be exploited by bourgeois manufacturers. These individuals were merely responding to a rigid conceptual framework by attending to the capacities of empirical reality in their efforts to cultivate practical knowledge. As Siegfried Giedion reminds us, "Mechanization is neutral. What matters is how one uses mechanization."[79] That industrial capitalism would repurpose the contributions of eighteenth-century aesthetic discourse is clear, but this process would not be absolute. Despite the violent legacies of industrial technologies like Watt's steam engine, the discursive formations embedded within such machines remained persistently malleable, capable of disrupting the very systems they constituted.

NOTES

1. James Watt, Fire and Steam Engines, &c., U.K. Patent 1432, issued April 28, 1784.

2. Karl Marx, *Capital*, vol. 1, trans. Ben Fowkes (London: Penguin, 1990), 499.

3. Paul Crutzen and Eugene Stoermer, "The 'Anthropocene,'" *IGBP Newsletter* 41 (2000): 17–18. The specific dating of the Anthropocene has, of course, become a major point of contention among humanists and social scientists. For further discussion of the Anthropocene starting date, see Jeremy Davies, *The Birth of the Anthropocene* (Berkeley: University of California Press, 2016), 41–68; Simon L. Lewis and Mark A. Maslin, *The Human Planet: How We Created the Anthropocene* (New Haven, CT: Yale University Press, 2018); Christophe Bonneuil, "The Geological Turn: Narratives of the Anthropocene," in *The Anthropocene and the Global Environmental Crisis: Rethinking Modernity in a New Epoch*, ed. Clive Hamilton, Christophe Bonneuil, and François Gemenne (Abingdon, UK: Routledge, 2015), 17–31; J. R. McNeill and Peter Engelke, *The Great Acceleration: An Environmental History of the Anthropocene since 1945* (Cambridge MA: Harvard University Press, 2016); Bruce D. Smith and Melinda A. Zeder, "The Onset of the Anthropocene," *Anthropocene* 4 (2013): 8–13.

4. Timothy Morton, *Hyperobjects: Philosophy and Ecology after the End of the World* (Minneapolis: University of Minnesota Press, 2013), 7.

5. Marxist historiography has operated on similar assumptions. G. A. Cohen, for example, identifies the locus of industrial capitalist society in "technical knowledge" above and beyond the physical embodiment of the machine itself. As Cohen argues, "Destroy all steam engines but preserve knowledge of how to make and use them and, with a bit of luck in the matter of raw materials, you can soon return to the status quo ante. Destroy the knowledge and preserve the engine and you have a useless ensemble of metal, a material surd, a relic of the future." Cohen, *Karl Marx's Theory of History: A Defence* (Princeton, NJ: Princeton University Press, 2000), 41.

6. Alan Mikhail, "Enlightenment Anthropocene," *Eighteenth-Century Studies* 49, no. 2 (Winter 2016), 211–231.

7. Marx, *Capital*, 499.

8. John Bourne, *A Treatise on the Steam Engine in Its Application to Mines, Mills, Steam Navigation, and Railways* (London, 1846), 211.

9. See Bonnie Blackwell, "The Pleasures of Geometry: Math and Melancholia in Rousseau's *La Nouvelle Helóïse*," *European Romantic Review* 15, no. 1 (March 2004): 89–112; Laura Søvsø Thomasen and Henrik Kragh Sørensen, "The Irony of Romantic Mathematics: Bridging the Historiographies of Literature and Mathematics," *Configurations* 24, no. 1 (Winter 2016): 53–70; Matthew Wickman, *Literature after Euclid: The Geometric Imagination in the Long Scottish Enlightenment* (Philadelphia: University of Pennsylvania Press, 2016).

10. For more on the history of the serpentine line, see James Bunn, *Wave Forms: A Natural Syntax for Rhythmic Language* (Stanford, CA: Stanford University Press, 2002); Miranda Stanyon, "Serpentine Sighs: De Quincey's *Suspiria de Profundis* and the Serpentine Line,"

Studies in Romanticism 53, no. 1 (Spring 2014): 31–58; David Summers, "*Maniera* and Movement: The *Figura Serpentinata*," *Art Quarterly* 35 (1972): 269–273.

11. See Francis Hutcheson, *An Inquiry into the Original of Our Ideas of Beauty and Virtue* (London, 1726). For a discussion of the mathematical nature of the "Shaftesbury-Hutcheson Beautiful," see Ronald Paulson, *Hogarth*, vol. 3, *Art and Politics, 1750–1764* (New Brunswick, NJ: Rutgers University Press, 1993), 65–76.

12. Tim Ingold, *Lines: A Brief History* (London: Routledge, 2007), 152.

13. W.J.T. Mitchell, *Articulate Images: The Sister Arts from Hogarth to Tennyson* (Minneapolis: University of Minnesota Press, 1983), 127.

14. William Shenstone, "Verses Written in the Garden of William Shenstone, Esq.," in *The Poetical Works of William Shenstone*, vol. 1 (London, 1780), 58.

15. Siegfried Giedion, *Mechanization Takes Command: A Contribution to Anonymous History* (Minneapolis: University of Minnesota Press, 2013), 312.

16. Quoted in Ingold, *Lines*, 159.

17. See Tom Turner, *English Garden Design: History and Styles since 1650* (Suffolk, UK: Antique Collectors' Club, 1986).

18. William Hogarth, *The Analysis of Beauty* (New Haven, CT: Yale University Press, 1997), 65; hereafter cited parenthetically in the text.

19. See Rosamaria Loretelli, "The Aesthetics of Empiricism and the Origin of the Novel," *The Eighteenth Century* 41, no. 2 (Summer 2000): 83–109.

20. Michel Baridon, "Hogarth's 'Living Machines of Nature' and the Theorisation of Aesthetics," in *Hogarth: Representing Nature's Machines*, ed. David Bindman, Frédéric Ogée, and Peter Wagner (Manchester: Manchester University Press, 2001), 92.

21. Ibid., 94.

22. For a discussion of Hogarth's influence on Rouquet, see Paulson, *Hogarth*, 193–196.

23. Jean Andre Rouquet, *The Present State of the Arts in England* (London, 1755), 7.

24. Ibid., 9.

25. Scott R. MacKenzie, "My Chapter upon Lines: Motion, Deviation, and Lineation in Eighteenth-Century British Aesthetics," *Criticism* 61, no. 1 (2019): 6.

26. Timothy Erwin, "William Hogarth and the Aesthetics of Nationalism," *Huntington Library Quarterly* 64, no. 3 (2001): 393.

27. Paulson, *Hogarth*, 97.

28. Mitchell, *Articulate Images*, 131; Frédéric Ogée, "From Text to Image: William Hogarth and the Emergence of a Visual Culture in Eighteenth-Century England," in Bindman, Ogée, and Wagner, *Hogarth*, 10.

29. Paulson, *Hogarth*, 73.

30. See Anne Puetz, "Design Instructions for Artisans in Eighteenth-Century Britain," *Journal of Design History* 12, no. 3 (1999): 217–239.

31. Paulson, *Hogarth*, 122.

32. "Defence of Coach-Building, and a Few Words on Steam-Carriages," *Mechanics Magazine* 18 (1833): 209.

33. Quoted in Ian Carter, *Railways and Culture in Britain: The Epitome of Modernity* (Manchester: Manchester University Press, 2001), 264.

34. Ruth Mack, "Hogarth's Practical Aesthetics," in *Mind, Body, Motion, Matter: Eighteenth-Century British and French Literary Perspectives*, ed. Mary Helen McMurray and Alison Conway (Toronto: University of Toronto Press, 2016), 22, 29. See also Abigail Zitin, *Practical Form: Abstraction, Technique, and Beauty in Eighteenth-Century Aesthetics* (New Haven, CT: Yale University Press, 2020).

35. Martin Heidegger, "The Question Concerning Technology," in *The Question Concerning Technology and Other Essays*, ed. and trans. William Lovitt (New York: Garland Publishing, 1977), 14.

36. Ibid., 14, 16.

37. Liliane Hilaire-Pérez and Catherine Verna, "Dissemination of Technical Knowledge in the Middle Ages and the Early Modern Era," *Technology and Culture* 47 (July 2006): 539.

38. Pamela H. Smith, "Art, Science, and Visual Culture in Early Modern Europe," *Isis* 97 (March 2006): 95. See also Smith, *The Body of the Artisan: Art and Experience in the Scientific Revolution* (Chicago: University of Chicago Press, 2004).

39. Dániel Margócsy, *Commercial Visions: Science, Trade, and Visual Culture in the Dutch Golden Age* (Chicago: University of Chicago Press, 2014); Eileen Reeves, *Painting the Heavens: Art and Science in the Age of Galileo* (Princeton, NJ: Princeton University Press, 1997). See also Anita Guerrini, *The Courtiers' Anatomists: Animals and Humans in Louis XIV's Paris* (Chicago: University of Chicago Press, 2015); Harold J. Cook, *Matters of Exchange: Commerce, Medicine, and Science in the Dutch Golden Age* (New Haven, CT: Yale University Press, 2007).

40. Ben Russell, *James Watt: Making the World Anew* (London: Reaktion, 2014), 197, 201.

41. See Peter Brooks, *Reading for the Plot: Design and Intention in Narrative* (New York: Knopf, 1984), 41–44; Tamara S. Ketabgian, *The Lives of Machines: The Industrial Imaginary in Victorian Literature and Culture* (Ann Arbor: University of Michigan Press, 2011); Andreas Malm, *Fossil Capital: The Rise of Steam Power and the Roots of Global Warming* (London: Verso, 2016), 194–222; Anson Rabinbach, *The Human Motor: Energy, Fatigue, and the Origins of Modernity* (Berkeley: University of California Press, 1992); Russell, *James Watt*, 170–202.

42. Russell, *James Watt*, 9.

43. I refer to Watt's linkage and the parallel motion device he developed shortly after the 1784 patent was issued interchangeably throughout. The parallel motion device used the exact same mechanism that we see in the 1784 patent, but it added a fourth bar to the contraption to save space.

44. For more on the technical history of Watt's linkage, see Ben Marsden, *Watt's Perfect Engine: Steam and the Age of Invention* (New York: Columbia University Press, 2002); James Patrick Muirhead, *The Origin and Progress of the Mechanical Inventions of James Watt* (London: John Murray, 1854); William Rosen, *The Most Powerful Idea in the World: A Story of Steam, Industry, and Invention* (New York: Random House, 2010).

45. Helmut Müller-Sievers, *The Cylinder: Kinematics of the Nineteenth Century* (Berkeley: University of California Press, 2012), 43.

46. François Arago, *Historical Eloge of James Watt*, trans. James Patrick Muirhead (London: John Murray, 1839), 85.

47. Ibid., 214.

48. Russell, *James Watt*, 167.

49. Robert Stuart, *A Descriptive History of the Steam Engine* (London: Knight and Lacey, 1824), 190–191.

50. Michel Serres, "Turner Translates Carnot," in *Hermes: Literature, Science, Philosophy*, ed. and trans. Josué V. Harari and David F. Bell (Baltimore: Johns Hopkins University Press, 1982), 56.

51. Ibid., 56.

52. Matthew Boulton to James Watt, July 17, 1784, Boulton and Watt Collection, MS 3147/3/8/10, Library of Birmingham, UK.

53. Watt, Fire and Steam Engines &c.

54. Quoted in David Philip Miller, *James Watt, Chemist: Understanding the Origins of the Steam Age* (London: Pickering and Chatto, 2009), 97.

55. Müller-Sievers, *The Cylinder*, 29.

56. Ibid.

57. Jesse Oak Taylor, "The Novel after Nature, Nature after the Novel: Richard Jefferies's Anthropocene Romance," *Studies in the Novel* 50, no. 1 (Spring 2018): 110.

58. Ibid.

59. Others have offered similar arguments about the ineptitude of traditional literary genres in responding to the ongoing environmental crisis. See Amitav Ghosh, *The Great Derangement: Climate Change and the Unthinkable* (Chicago: University of Chicago Press, 2016); Ursula K. Heise, "Lost Dogs, Last Birds, and Listed Species: Cultures of Extinction," *Configurations* 18, nos. 1–2 (Winter 2010): 49–72; Stephanie LeMenager, "Climate Change and

the Struggle for Genre," in *Anthropocene Reading: Literary History in Geologic Times*, ed. Tobias Menely and Jesse Oak Taylor (University Park: Pennsylvania State University Press, 2017), 220–238.

60. See Rebecca Tierney-Hynes, "Fictional Mechanics: Haywood, Reading, and the Passions," *The Eighteenth Century* 51, nos. 1–2 (Spring–Summer 2010): 153–172.

61. Laurence Sterne, *The Life and Opinions of Tristram Shandy, a Gentleman* (London: Penguin, 2003), 257.

62. Ibid., 253.

63. Joseph Drury, *Novel Machines: Technology and Narrative Form in Enlightenment Britain* (Oxford: Oxford University Press, 2017), 138.

64. Sterne, *Tristram Shandy*, 64.

65. Ibid., 63–64.

66. Viktor Shklovsky, *Theory of Prose*, trans. Benjamin Sher (Champaign, IL: Dalkey Archive, 1990), 165.

67. Christina Lupton, "Contingency, Codex, the Eighteenth-Century Novel," *ELH* 81, no. 4 (Winter 2014): 1180. See also Eric Rothstein, *Systems of Order and Inquiry in Later Eighteenth-Century Fiction* (Berkeley: University of California Press, 1975).

68. Sterne, *Tristram Shandy*, 425.

69. For a discussion of the competing representational systems depicted in the Shandy household, see Alex Solomon, "The Novel and the Bowling Green: Toby Shandy's Diagrammatic Realism," *Philological Quarterly* 95, no. 2 (2016): 269–291.

70. Sterne, *Tristram Shandy*, 323–324.

71. Drury, *Novel Machines*, 4.

72. Roy C. Caldwell, "*Tristram Shandy*, Bachelor Machine," *The Eighteenth Century* 34, no. 2 (Summer 1993): 106. See also Judith Hawley, "*Tristram Shandy*, Learned Wit, and Enlightenment Knowledge," in *The Cambridge Companion to Laurence Sterne*, ed. Thomas Keymer (Cambridge: Cambridge University Press, 2009), 34–48.

73. See also Peter Conrad, *Shandyism: The Character of Romantic Irony* (Oxford: Basil Blackwell, 1978); Martin Meisel, *Chaos Imagined: Literature, Art, Science* (New York: Columbia University Press, 2016), 307–309.

74. Sterne, *Tristram Shandy*, 588.

75. Ibid., 334.

76. See Paul Slack, *The Invention of Improvement: Information and Material Progress in Seventeenth-Century England* (Oxford: Oxford University Press, 2015).

77. Sterne, *Tristram Shandy*, 510.

78. Gilbert Simondon, *On the Mode of Existence of Technical Objects*, trans. Cécile Malaspina and John Rogove (Minneapolis: Univocal, 2017), 18.

79. Giedion, *Mechanization Takes Command*, 345.

CHAPTER 6

Technology, Temporality, and Queer Form in Horace Walpole's Gothic

Emily M. West

In *A Description of the Villa . . . at Strawberry-Hill*, Horace Walpole guides readers through his faux-medieval gothic mansion and its collections in a virtual version of the tours visitors could take through his home. Entering the building in person or in text, such visitors encounter a curious light: "You first enter a small gloomy hall paved with hexagon tyles, and lighted by two narrow windows of painted glass, representing St. John and St. Francis. This hall is united with the staircase, and both are hung with gothic paper. . . . In the well of the staircase, by a cord of black and yellow, hangs a gothic lanthorn of tin japanned, designed by Mr. Bentley, and filled with painted glass; the door of it has an old pane with the arms of Vere earl of Oxford."[1] From its position in the stairwell, this gothic lantern produced the signature effect of Strawberry Hill's "venerable barbarism": the "gloomth of abbeys and cathedrals" that Walpole cultivated in his home.[2] Illuminating (barely) the space that Walpole named "the most particular and chief beauty of the castle"—a space that served as the staging ground for the dream that inspired *The Castle of Otranto*—the light created its gothic atmosphere (see figures 6.1–6.4).[3]

The *Description of the Villa . . . at Strawberry-Hill* and its tour of the home that Walpole called "a paper Fabric" suggest how textual and material forms collaborate to produce the gothic genre and its effects.[4] In this chapter, I explore one aspect of that collaboration by analyzing Strawberry Hill's lantern as a gothic technology. Through its operation, the lantern materializes gothic forms, and attending to its materiality offers new insight on how those forms remade normative models of temporality and sexuality across Walpole's multidisciplinary work. Indeed, a key claim of this essay is that temporality and sexuality are inextricably bound together in Walpole's gothic. Here my argument builds on foundational queer theoretical analyses of Walpole's work (by George Haggerty, Jill

Figure 6.1. The "gothic lanthorn" (ca. 1753–1755), designed by Horace Walpole and Richard Bentley and constructed by William Hallett. The lantern is currently held by the Lewis Walpole Library, Yale University, where it hangs in the executive director's office. A replica of the lantern hangs in its original position at Strawberry Hill. Image courtesy of the Lewis Walpole Library, Yale University.

Figure 6.2. The lantern in operation in the executive director's office at the Lewis Walpole Library. Photo by the author.

Campbell, Matthew Reeve, and others) by using queer temporality as my primary framework. Drawing on the queer theories of time articulated by scholars such as Elizabeth Freeman, Jack Halberstam, and Lee Edelman, I show how Walpole's irreverently critical treatment of both linear historicity and the heterosexual family act as linked disruptions of what Halberstam names the "time of inheritance": a conceptual structure in which the heteropatriarchal family unit organizes a linear model of history and futurity and through which normative systems are endlessly reproduced.[5] By "stepping out of the linearity of straight time," Walpole's gothic ultimately (and unexpectedly) enacts the kind of queer methodology of hope detailed by José Esteban Muñoz in *Cruising Utopia*.[6] As part of this methodology, the lantern materializes the formal aesthetics of explosion, fragmentation, mingling, and remaking that likewise shape Walpole's gothic antimarriage plots in *The Castle of Otranto* and *The Mysterious Mother*. Further, as a technology of queer time, the lantern revises the way that technological objects were called on across the eighteenth century to manifest Enlightenment modernity as a progression from the superseded past. By generating

Figure 6.3. The replica lantern in operation at Strawberry Hill. Photo by the author.

gloomth instead of enlightenment, Walpole's gothic technology suggests other ways of making a future.

To think about how that future is materialized, I begin with the gothic lantern's material form. Walpole expands the account he gives of the light in the *Description* in a letter to his friend George Montagu: "I have filled Mr Bentley's Gothic lanthorn with painted glass, which casts the most venerable gloom on the stairs that was ever seen since the days of Abelard. The lanthorn itself in which I have stuck a coat of the Veres is supposed to have come from Castle Henningham."[7] The painted glass came from the assortment that Walpole had amassed as one of England's earliest and most avid collectors of decorative glass.[8] The largest part of his collection was early modern and Dutch, but it also included "medieval armorials, shields, and badges," Tudor artifacts, and newly commissioned pieces.[9] The lantern, like Strawberry Hill's other assemblages of decorative glass, combined pieces from these disparate locations and historical periods with "eclectic abandon," but in Walpole's descriptions of the fixture, he consistently and uncharacteristically highlights the Vere family coat of arms incorporated into its collaged arrangement (see figures 6.5 and 6.6).[10] The Veres had been of interest to Walpole since he visited their estate at Hanworth and ancestral castle at Hedingham (which he called Henningham) in the 1740s and, declaring himself "Vere-mad," threw himself into researching the family's genealogy.[11] This was a history that stretched back to the Norman invasion: Aubrey de Vere, who established Hedingham, was granted his title and lands by William the Conqueror. Thomas

Figure 6.4. Richard Bentley, "Perspective of the Hall & Staircase at Strawberryhill," from *Drawings and Designs by Richd. Bentley*, Lewis Walpole Library, folio 49 3585, leaf 30a. This image looks down on the hall from a partial ascent of the staircase. The lantern, not included in this illustration, hung in the middle of this space. Image courtesy of the Lewis Walpole Library, Yale University.

Figure 6.5. Photograph of the gothic lantern showing the side in which the Vere coat of arms is inserted. Image courtesy of the Lewis Walpole Library, Yale University.

Figure 6.6. George Perfect Harding, "The Lanthron [*sic*] in the Red Hall," an unfinished drawing of the gothic lantern in which only the Vere arms are distinguished among the pieces of stained and painted glass. From Thomas Kirgate's extra-illustrated copy of *A Description of the Villa . . . at Strawberry Hill*, Lewis Walpole Library, quarto 33 30, copy 6. Image courtesy of the Lewis Walpole Library, Yale University.

Babington Macaulay, in his epochal Whig history of England, names the Veres "the longest and most illustrious line of nobles that England has seen."[12] Hedingham was likewise, as its eighteenth-century tenant Lewis Majendie declared, "a structure of no common note, having been the chief residence annext to an estate belonging by grant from the Conqueror to the very antient and noble family of De Vere, Earls of Oxford; in which the family remained, from those early times, almost without interruption, till the year 1655."[13]

The Vere family and castle are, in these accounts, living monuments of English history. The Vere lineage thus makes obvious the entanglement of family and state histories and futures, since the continuation of this family line is inseparable from the social reproduction of Englishness itself. This mode of entanglement has been analyzed by theorists of queer time, who trace how normative forms of temporality and sexuality are linked. Elizabeth Freeman, for example, explains how "the state and other institutions, including representational apparatuses, link properly temporalized bodies" to "teleological schemes of events or strategies for living such as marriage, accumulation of health and wealth for the future, reproduction," and "childrearing."[14] Through these schemes, bodies are "bound into socially meaningful embodiment through temporal regulation," and the future is, Lee Edelman has argued, imagined genealogically through the figure of the child who will inherit and reproduce them in turn.[15] Jack Halberstam shows how "inheritance" thus transcends the function of individual property transfer to become a figure for the "collective legacy from which a group will draw a properly political future" along national and racial lines. Inheritance as a form organizes a "generational time within which values, wealth, goods, and morals are passed through family ties from one generation to the next," connecting "the family to the historical past of the nation, and . . . the future of both familial and national stability."[16]

The Vere genealogy organizes the time of inheritance by articulating a long, illustrious line promising the perpetual continuation of a national futurity conceptualized and realized through heterosexual forms. This is, I argue, both why Walpole was so fascinated with the family's legacy and why he remade that legacy in key ways. One such method of remaking was textual, comprising accounts that puncture that legacy, as in this letter to Montagu:

> I have been to make a visit of two or three days to Nugent, and was carried to see the last remains of the glory of the old Aubrey de Veres, Earls of Oxford. They were once masters of almost this entire county, but quite reduced even before the extinction of their house; the last Earl's son died at a miserable cottage that I was shown at a distance, and I think another of the sisters, besides Lady Mary Vere, was forced to live upon her beauty. Henningham Castle, where Harry the Seventh was so sumptuously banqueted and imposed that villainous fine for his entertainment, is now shrunk to one vast ruinous tower.[17]

Whereas Macaulay and Majendie emphasize dynastic glory, Walpole foregrounds decay and extinction, his account narrating the ruin of family and seat. It is this ruin that makes available to Walpole a piece of the Veres' remains in the form of the lantern, a fictional artifact from Hedingham that may or may not contain an actual fragment of stained glass from the castle. Rather than the veracity or provenance of the painted Vere coat of arms, what is most relevant here is how the glass pane's form interrupts the time of inheritance through its visible dislocation and fragmentation of the family's legacy. The formal logic that enables the fragment's appropriation in the lantern is familiar from Walpole's gothic fictions: that of explosion.

Explosion is introduced in Walpole's literary work through its relationship to time and genre: *The Castle of Otranto*'s first preface (written to introduce the book to the public as a medieval story recovered from "the library of an ancient Catholic family in the north of England") pre-emptively atones for the book's excesses with an "apology" for its inclusion of "Miracles, visions, necromancy, dreams, and other preternatural events" that "are exploded now even from romances."[18] The preface here employs a familiar timescale of development in which past forms are "rejected, discontinued" and "held in contempt" as part of a progression toward a rational and enlightened future.[19] Depending on your critical perspective, Walpole's attempt to "wind up the [reader's] feelings . . . till they [become] for a moment identified with those of a ruder age" either reinforces or rejects this timescale.[20] Regardless, such gothic encounters with the past must inevitably negotiate the injunction to "Remember the country and the age in which we live. Remember that we are English, that we are Christians."[21] What interests me here is the way that exploded forms in *Otranto* link these national timescales to their necessary realization through the heterosexual family. This link is made in the book's first pages when Otranto's tyrannical patriarch, Manfred, sees Conrad—his sole, infirm male heir—crushed to death by a heraldic helmet that materializes spectacularly out of time and place. As Walpole narrates, "What a sight for a father's eyes!—He beheld his child dashed to pieces, and almost buried under an enormous helmet" (18). In this moment, explosion's function to "eject" and "cast out" past forms in a progression to futurity meets its material connotations of "burst[ing], shatter[ing], or [breaking] apart violently" in a manner that "typically scatter[s] fragments outwards,"[22] as Otranto's lineal future is smashed into "bleeding mangled remains" (19) before Conrad can consummate a dynastic marriage. The juxtaposition between exploded narrative and embodied forms staged through *Otranto*'s preface and opening pages suggests that in Walpole's gothic what is ejected or cast out are not superseded past forms but instead the future that a genealogical progression from such forms insists into being. Rather than the line, the operative structure in this queer gothic relation to past and future is the fragment.

The Vere pane, as the dislocated fragment of a shattered dynastic legacy, evinces the same material form and narrative relation to past and future that shape *Otranto*'s dissident temporalities. Reading the gothic lantern alongside Walpole's gothic texts is generative because it highlights what this explosive, fragmentary relation to past and future can *make*, rather than simply what it rejects, interrupts, or destroys. Here it is helpful to consider both how the lantern was made and what it makes. To begin with its construction, as a centerpiece in Strawberry Hill, the lantern was assembled in the same collective mode as the rest of the house. George Haggerty's and Jill Campbell's germinal work has traced how Strawberry Hill acted as a locus for Walpole's intense relations with his closest friends, who became collaborators in this space of fantasy.[23] The "more intimate a relationship a friend shared with Walpole," Haggerty observes, "the more deeply he would implicate that friend in this Gothic product of his imagination."[24] Or, as Eugenia Zuroski has recently put it, "In the Strawberry Hill project, Walpole contrived a way to dream together with his friends as a matter of quotidian practice."[25] This collective of friends replaced the dynastic family that such a home conventionally commemorated. The gothic lantern is most directly the product of Walpole's collaboration with Richard Bentley, with whom he designed the fixture, along with a number of Strawberry Hill's spaces and artifacts.[26] The "japanned" lantern's chinoiserie elements mark it as the invention of this intimate relation. During the time of its creation Walpole rejected chinoiserie forms, but Bentley remained enamored of Asian aesthetics and continued to propose their incorporation into Strawberry Hill and other shared projects. Because Walpole scorned the mingling of gothic and chinoiserie aesthetics as an "unnatural copulation," their conjunction in the japanned gothic lantern signals that this object is a collaborative thing: the product of a "practice of intimacy that accommodates difference of taste through assertions of affection."[27] The lantern is visibly the creation of two men's shared dream. Stained and painted glass also functioned as an intimate site in Walpole's inner circle, as expressed through the descriptions of Strawberry's glass shared in his letters and in requests to his friends for more. Walpole writes to Horace Mann a few years before the lantern's construction that he is giving him a "commission": "If you can pick me up any fragments of old painted glass, arms or anything, I shall be excessively obliged to you."[28] These "historic bits," loosed from ancestral estates, now circulate through and signal the queer forms of relation that made Strawberry Hill.[29]

Queer forms likewise shape these fragments' arrangement in the lantern. Critics have commented on the unruly irreverence that characterizes Walpole's use of disparate glass pieces throughout Strawberry Hill, the way he "combines them in an almost sacrilegious manner to produce a decorative melange."[30] This discontinuous arrangement recalls the patchwork aesthetic that Alice Tweedy McGrath has analyzed in Jane Barker's writing as a queer method that resists continuity and closure: as she writes, "Patchwork form is governed by juxtaposition rather than cohesion, patchwork logic by contiguity rather than cause and effect."[31] Complementing the content of a narrative that "depicts illicit desires that thwart the primacy of heterosexual marriage," Barker's *A Patch-Work Screen for the Ladies* uses a patchwork method to organize "stories without reference to the telos of the marriage plot or any other endpoint."[32] Likewise, Walpole's collaged arrangements of glass include heraldic fragments like the coat of arms in ways that evade restoring the genealogical trajectories that such artifacts are meant to generate, embedding these pieces instead in antiteleological assemblages.

The discontinuous queer aesthetic that determines the lantern's construction also structures what it makes: namely, the gloomth produced when the lantern is lit. In its imagined place at the Veres' ancestral seat, the illuminated coat of arms would literalize the concept of lineage by making lustrous "the longest and most illustrious line of nobles that England has seen," continually casting the Veres forward to a future conceptualized and materialized through the image of family.[33] As a repurposed fragment amid an array of other disparate glass pieces in the lantern (a motley collection of religious iconography, natural history, and other heraldic symbols from various sources), it is instead lit into a haze of colored light, the family crest mixing promiscuously with the images against which it is juxtaposed. As an optical technology, the lantern exploits stained glass's formal ambivalence, the way that it functions as "an indiscrete medium": "Unlike other visual media, which remain within the discrete boundary of their material frames," Anne F. Harris observes, "stained glass exceeds its materiality" and "dissolve[s] narrative into a kaleidoscopic array."[34] While the heraldic glass is designed to exceed the bounds of its materiality in order to enact the family line's projection into the future, here the image of the Veres instead dissolves into an atmosphere of what Walpole called "conventual gloom."[35] As Matthew Reeve has traced, this "conventual" ambiance is part of the "rich tradition during the eighteenth century in Protestant England that considered Catholicism to be a mode of religious transgression that coded sexual transgression."[36] While Walpole's evocations of Catholicism skirted the deliberate outrageousness of the "Monks of Medmenham" (a hellfire club that, in the 1750s, performed rituals of sexual debauchery with sex worker "nuns" at Medmenham Abbey),[37] "popery" was central to the fantasy he cultivated at Strawberry Hill. He wrote to William Cole that "I like it as I do chivalry and romance. They all furnish one with ideas and visions, which Presbyterianism does not. A Gothic church or convent fill one with romantic dreams."[38] In this letter and elsewhere, Reeve shows, Walpole's "fantasies inspired by Gothic architecture" are suffused with "an erotic charge."[39] In the lantern that produced "the gloomth of abbeys and cathedrals" in which visitors were enveloped on entering Strawberry Hill, the Vere crest dissipates into a nimbus of charged reverie.[40]

As I suggested earlier, what particularly interests me about the lantern and the gloomth it projects are their implications for understanding the gothic's relationship to the past and the future as generative rather than only disruptive or destructive. Of course, explosion (the Walpolean gothic's method of shattering genealogical relations between past and future) is clearly crucial to the lantern's function. Illuminating the entrance to Strawberry Hill, its coat of arms is a fragment not only out of place but also out of the genealogical continuity premised on and realized through that place; the ancestry it casts forth has no continuance except as a sign of its own obsolescence. Like Conrad's "disfigured corpse" (*Otranto*, 19), the Vere remains here function as an emblem of extinction rather than a legacy. Yet as it refuses the time of inheritance, this gothic technology also makes something: a fantasy materialized in the real world, one constructed from and for queer intimacies. In this way the lantern generates *possibility*, what Muñoz describes in *Cruising Utopia* as "the illumination of a horizon of existence," of "a queerness to come."[41] In that text, Muñoz proposes that "queerness is essentially about the rejection of a here and now and an insistence on potentiality or concrete possibility for another world" and outlines how "we can glimpse the worlds proposed and promised by queerness in the realm of the aesthetic," which "frequently contains blueprints and schemata of a forward-dawning futurity."[42] To construct and witness such blueprints is to undertake a "queer utopian hermeneutic": a critical methodology of hope.[43] Considered in this light, Walpole's gothic may not actually be primarily about the past and its relation to the present but instead, as an aesthetic of queer utopian worldmaking, centrally about the future; to return to Muñoz, we could read it as "a backward glance that enacts a future vision."[44] In this way, though the gothic may seem like the most paranoid of genres,[45] we can recognize the gothic lantern's construction from "historic bits" as an essentially reparative project: "additive," "accretive," a collaborative work of "assembl[ing] and confer[ring] plenitude" on broken relics and remains to illuminate another way forward.[46]

What, then, of the story Walpole dreamed in this gloomth? *The Castle of Otranto*'s stated moral (introduced in the preface and recalled several times throughout the narrative) warns that "*the sins of the father are visited on their children to the third and fourth generation*" (6), evoking what Eve Kosofsky Sedgwick calls the "dogged, defensive narrative stiffness of a paranoid temporality" that "takes its shape from a generational narrative . . . characterized by a distinctly Oedipal regularity and repetitiveness: it happened to my father's father, it happened to my father, it is happening to me, it will happen to my son, and it will happen to my son's son."[47] Yet, as explored earlier, forms of generational continuity are mangled from the first page of the text, and throughout the narrative, its gothic episodes elaborate the aesthetics of queer time materialized in Strawberry Hill's gothic lantern. This is perhaps most obvious in the famous scene in which a portrait in Otranto's gallery "quit[s] its pannel, and descend[s] on the floor" (25): a moment that refuses linearity in at least three ways. First, the portrait of Manfred's grandfather Ricardo is meant to demonstrate his descendant's ancestral claim to Otranto, yet readers later learn that Ricardo murdered Alfonso, Otranto's lord, and forged a "fictitious will" naming himself as Alfonso's heir (104). Like the will, the painting underpins a false dynastic claim, suggesting both how genealogical forms function primarily to transmit power and property forward along normative lines and how they are ultimately fictions constructed to serve such ends. Further, the panel's painted surface is restless; like a lit stained-glass pane, it refuses to "remain within the discrete boundary" of its representational frame. Instead, quitting it—"exceed[ing] its materiality"[48]—the image performs the disruption of lineage and linearity as an ancestor appears out of time and place. Finally, the appearance of the panel in Walpole's novel is itself an anachronism. As a contemporary reviewer noted, "We cannot help thinking that this circumstance is some presumption that the castle of Otranto is a modern fabrick; for we doubt whether pictures were fastened in pannels before the year 1243."[49] As the reviewer suspects, Ricardo's painting actually references one in the gallery at Strawberry Hill; Walpole wrote to his friend Cole, "When you read of the picture quitting its panel, did not you recollect the portrait of Lord Falkland all in white in my gallery?"[50] Walpole's insertion of the seventeenth-century portrait of Falkland in his supposedly medieval story performs the same antilinearity as the fictional portrait's animation, using a method of juxtaposition and anticontinuity transposed from material to narrative gothic forms. Walpole's reference to the Falkland portrait also links *Otranto*'s unsettled ancestor to the ways Walpole deployed picture hangs and gallery walks at Strawberry Hill to articulate the "complex bonds that [he] shared with his homosocial circle," such that conventional dynastic practices of "collecting, arranging, displaying, and exchanging art" became "a venue for creating queer family romances" rather than a method of consolidating heteropatriarchal histories and futures.[51] The vision of a fragmented and ultimately futureless dynastic line emblematized by Ricardo's step beyond the historical frame interrupts the time of inheritance but does so in a way that evokes and extends Strawberry Hill's queer worldmaking.

These aesthetics of anachronism and antilineage are not confined to Manfred's illegitimate dynastic line but also characterize Otranto's legitimate heir, Theodore, who is himself an ancestral portrait roaming beyond the bounds of his frame. Theodore's "exact resemblance" (51) to "the picture of the good Alfonso in the gallery" (38) is marked by all who are familiar with the painting. Matilda fears that she dreams when she sees him because of it (51); and when Manfred sees Theodore in armor for the first time, he takes him for a "dreadful spectre" (using the same language he applied to Ricardo's ghost), finally declaring the youth a "ghastly phantom" that "unhinge[s] the soul" (76–77). Theodore, as the "very image" (81) of Alfonso's portrait, exists as both a material body and an immaterial "shade" (87), at once the living claimant to the "sceptre [Manfred] had received from [his] ancestors" (63) and the "dreadful spectre" of an ancestor out of time (76). These descriptions of Theodore as a spectral painted surface merge him with (rather than distinguish him from) Manfred's illegitimate, extinguished family line, since the same narrative and aesthetic forms characterize the manifestations of both ancestries. Extending this enmeshment of the two dynastic lines, Theodore is repeatedly taken by Isabella, Matilda, and Matilda's maid Bianca for *Conrad*'s ghost throughout the text, as Jason Farr has pointed out. Farr's queer disability studies analysis shows how this recurring confusion between Theodore and the infirm Conrad enacts an "uncanny crip troubling of the masculine ideal," and he examines one such instance in which Bianca interprets Theodore's presence as the ghost of Conrad's "young tutor, [who] drowned himself," and "the young Prince's" ghost "now met in the chamber below."[52] As Farr notes, the fantasy of "ghostly cohabitation in the chamber below is suggestive of a pederastic relationship," but the episode's queerness also operates at the level of form: Conrad's ghost "mingles with that of his deceased tutor" and with Theodore's living body as distinctions between individual figures dissolve into an atmosphere both haunting and charged with eroticism.[53]

This dissolution continues across the book, in episodes such as the one in which Theodore's "human form" is believed by Isabella to be "the ghost of her betrothed Conrad" (27). With these repeated dissolutions of the form meant to project Otranto's dynastic line into the future, Walpole's novel deploys the formal strategies that shape Strawberry Hill's gothic lantern and its operation as a technology of queer time. Just as the lantern's stained-glass collage "dissolve[s] narrative into a kaleidoscopic array," Theodore merges with his haunting double, blurring the genealogical line that he is supposed to materialize.[54] And, though Theodore closes the narrative by assuming his position as "the true prince of Otranto" and Isabella's husband, *Otranto*'s ending refuses to resolve into heterosexual futurity. Theodore is lord of an exploded castle ("The moment Theodore appeared, the walls of the castle behind Manfred were thrown down with a mighty force" [103]) and a dynastic marriage characterized primarily by sorrow rather than sexual union ("he was persuaded he could know no happiness but in the society of one with whom he could forever indulge the melancholy that had taken possession of his soul" [105]). *Otranto*'s conclusion deconstructs the material and social forms meant to project a heteropatriarchal future from the marriage plot's happy ending, leaving us instead with ruined fragments and an atmosphere of gloom. The ending thus formally mimics the entrance to Strawberry Hill and suggests how an ending might also be an entrance: how, in Muñoz's words, "a rejection of the here and now" can function as "an insistence on potentiality or concrete possibility for another world."[55]

Embracing gloomth's potentialities means rejecting Enlightenment futures to imagine something or somewhere else. This becomes clearer through Sean Silver's reading of the stairwell at Strawberry Hill (the staging ground both for the lantern and for the dream that inspired *The Castle of Otranto*). Silver identifies the staircase's landing as "perhaps the most important" of the vantage points in the house because of its challenge to eighteenth-century Whig visions of history invested in "rediscovering, at every point, a ubiquitous and gradual progression from 'darkness' to 'light'" (see figure 6.7).[56] As Silver observes, the stairwell's landing "stages and personalizes the central trope of Enlightenment historiography; the visitor stands at the moment between (as Walpole puts it in *Historic Doubts*) 'barbarity' and 'polish,' or (as Walpole puts it in his *Description*) 'gloom' and 'illumination.'"[57] Silver concentrates on how the position of Walpole's treasured suit of armor in this space performs a revisionist historiography: as a "Gothic artifact coming to light before one's eyes," it enacts a history in which the present, rather than existing as a "clean and well-lighted place of Enlightenment rationality," is "crisscrossed and sometimes traumatized by objects of the historical past."[58] Silver's reading thus emphasizes the disruptive effect of such exploded fragments and outmoded forms. Yet the gothic lantern also hangs "in the well of the staircase" directly beside the landing that Silver discusses, generating a visual effect that does not resolve into either light or darkness but instead mingles the two seemingly opposite qualities.[59] While the suit of armor interrupts a path to enlightenment, the lamp brings darkness and light together in a haze that pervades and envelops instead of advancing. As a technology of queer time, the lantern produces gloomth rather than enlightenment, materializing "the warm illumination of a horizon imbued with potentiality" instead of a linear progression from past to future.[60] The lantern's role in generating this "anticipatory illumination" is particularly signifi-

Figure 6.7. Edward Edwards, "Staircase at Strawberry Hill," 1784, from Walpole's extra-illustrated copy of *A Description of the Villa . . . at Strawberry-Hill*, Lewis Walpole Library, folio 49 3582. This image shows the view from the staircase's landing. The lantern, not included in this illustration, hung beside the landing. Image courtesy of the Lewis Walpole Library, Yale University.

cant given the way that technological objects were called on to manifest enlightenment.[61] For instance, Francis Bacon contrasts "the life of men in any of the most civilised provinces of Europe and in the most savage and barbarous region of New India." Bacon claims that their conditions "differ so much that deservedly it may be said that 'man is a God to man'" and that this difference can be attributed primarily to "Arts." He continues:

> It helps to notice the force, power and consequences of discoveries, which appear at their clearest in three things that were unknown to antiquity, . . . namely the art of printing, gunpowder and the nautical compass. In fact these three things have changed the face and condition of things all over the globe: the first in literature; the second in the art of war; the third in navigation; and innumerable changes have followed; so that no empire or sect or star seems to have exercised a greater power and influence on human affairs than those mechanical things.[62]

In Bacon's theorization, technological objects materialize modernity along with the timescale that situates modernity as a progression from a superseded past—a colonial timescale in which this past remains embodied by racialized subjects. Bacon's framework was later reprised in the writings of Royal Society fellows like Robert Hooke, who believed that "humane Nature" is "capable of being so far advanced by the helps of Art, and Experience, as to make some Men excel others in their Observations, and Deductions, almost as much as they do Beasts": an advancement that he believes "has been of late years accomplisht" by "the adding of *artificial Organs* to the *natural*" and more specifically through "the invention of Optical Glasses" like telescopes and microscopes.[63] For scientists like Hooke, such optical technologies served as both the engines and evidence of Enlightenment modernity.

Walpole's lantern is also, of course, an optical technology, and its function is not as distinct from Hooke's telescopes and microscopes as the narratives of material and social advancement associated with scientific instruments claim. In its projective function, the gothic lantern resembles the magic lantern, a device that illuminated and cast images from glass slides. Magic lanterns were popularized in the seventeenth century alongside technologies like the telescope and constructed using many of the same principles as these other optical mechanisms. Samuel Pepys writes in 1666 of an evening spent on "Opticke enquiries" with Richard Reeve, Restoration London's premier maker of optical technologies (including scientific instruments used by Christopher Wren, Henry Power, Robert Hooke, and Robert Boyle):[64] "He bringing me a frame with closes on, to see how the Rays of light do cut one another, and in a dark room with smoake, which is very pretty. He did also bring a lantern, with pictures in glass to make strange things appear on a wall, very pretty. We did also at night see Jupiter and his girdle and Satellites very fine with my 12-foot glass, but could not Saturne, he being very dark."[65]

Reeve's lantern is presented alongside the telescope, indicating their functional similarities: in fact, both had previously served as instruments of "natural magic" (as had the microscope). While microscopes and telescopes were remade as instru-

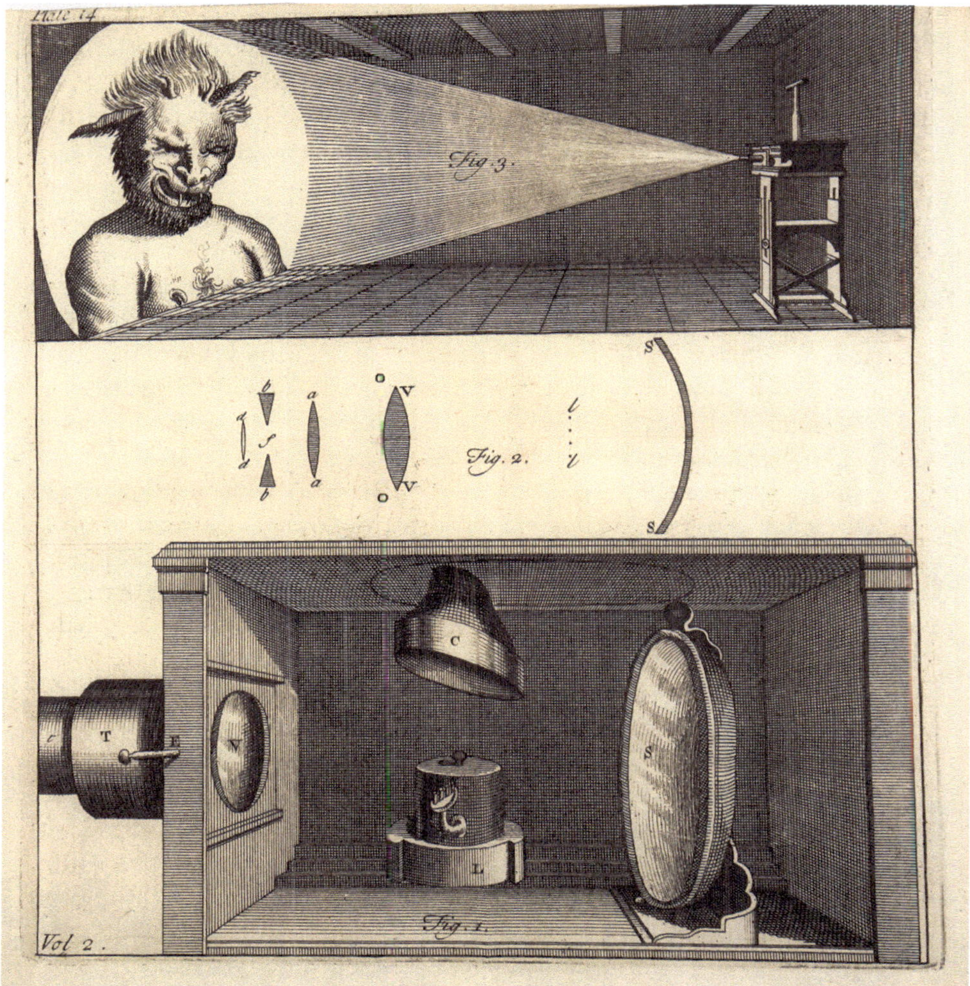

Figure 6.8. Willem 's Gravesande, *Mathematical Elements of Natural Philosophy, Confirm'd by Experiments; or, An introduction to Sir Isaac Newton's Philosophy* (1721), p. 105, via Wellcome Collection (Public Domain Mark).

ments of natural philosophy, the magic lantern "retained its association with natural magic longer than most instruments from the seventeenth century," as its name indicates.[66] Because of the lantern's persistent association with the occult, magic lantern shows troubled the narratives of progression to modernity materialized by the seemingly more rational mechanisms to which they were linked. In his analysis of the technology, Koen Vermeir observes that "Monsters appeared wherever a magic lantern was installed. Even [Willem] 's Gravesande's illustration of a magic lantern, drawn in a serious book on scientific demonstration lectures, projects a deformed monster" (see figure 6.8).[67] This inherent unruliness would eventually be capitalized on in the late eighteenth-century phenomenon of the phantasmagoria: "illusionistic exhibitions and public entertainments in which 'specters' were produced through the use of a magic lantern."[68] Like the magic lantern, Walpole's lantern produced visual effects that "gave the illusion of bridging the gap between the here and the over there, the real and the imagined, the intangible and the material."[69] Like the phantasmagoria, his lantern used forms meant to generate illustrious futures to cast instead specters roaming out of time and place. The gothic lantern thus participates in the magic lantern's troubling of the progressive historical narratives supposedly materialized by technological objects; and, crucially, Walpole's optical technology shows how these historical trajectories are bound up with the history of sexuality and the time of inheritance. Through the lantern's operation, Walpole brings together a critique of the heteropatriarchal family as a social form with a rejection of linear narratives of rational modernity and the mechanisms conscripted to produce Enlightenment futures, illuminating instead "a queerness to come."[70]

Walpole's gothic drama *The Mysterious Mother* plays with these optics of Enlightenment as its plot and form push *The Castle of Otranto*'s queer aesthetics to new extremities. Most significantly, *The Mysterious Mother* amplifies the threat of incest raised in *Otranto* by Manfred's marriage to Hippolita (he claims that they are "related . . . in the fourth degree" [48] in an attempt to legally dissolve the marriage) and sexual pursuit of his promised daughter-in-law, Isabella. In *The Mysterious Mother*, such genealogical perversity is elevated to the play's central secret and driving narrative force. In it, the Countess of Narbonne has played a bed trick to secretly sleep with her son, Edmund, and their daughter, Adeliza, is the product of the incestuous union. When Edmund returns to Narbonne sixteen years after his guilty mother banished him, still unaware of the events that led to his exile, he falls in love with Adeliza (whom the Countess has raised as an orphan ward) and marries her; the revelation of Adeliza's true parentage destroys the marriage and concludes the play.

Florian, Edmund's companion on his journey back to his ancestral estate, imagines the heir's return to Narbonne as a triumph of modernity. Rebuking the monk Benedict, who has held devious sway over the Countess during Edmund's exile, he describes

> These sixteen years . . . my friend Edmund pin'd in banishment:
> While masses, mummings, goblins and processions
> Usurp'd his heritage, and made of Narbonne
> A theatre of holy interludes
> And sainted frauds. But day darts on your spells.
> Th'enlighten'd age eschews your vile deceits,
> And truth shall do mankind and Edmund justice.[71]

In Florian's vision, Edmund's assumption of his rightful genealogical place at Narbonne enacts in miniature a historical progression from past to future, from primitive superstition to rationality: a progression conceptualized through a transition from darkness to light and materialized through Edmund's marriage (and future progeny). Edmund likewise uses illumination to imagine this future when he soliloquizes his wish to

> Make . . . the blooming Adeliza mine,
> And bear, of me unquestion'd, Narbonne's sceptre;
> Till life's expiring lamp by intervals
> Throws but fainter and a fainter flash,
> And then relumes its wasted oil no more. (69)

Edmund here uses the image of a lamp to express the length of an individual life; after the lamp is extinguished, that life can continue only dynastically, through future heirs to the Narbonne line. Yet the play makes a grim mockery of Edmund's words with the revelation of his double incest. As part of this revelation, his mother explains its circumstances: her husband, on whom her "luxurious fancy / Doated," had been absent for eighteen months; at his expected return she discovered instead his death, which raised a "storm of disappointed passions" that "Assail'd [her] reason, fever'd all [her] blood" (114–115). In this state, she tells her son, "my fancy saw thee / Thy father's image," and she secretly took Edmund's lover's place in a tryst (115).

This confusion between patriarch and heir recurs when the Countess sees Edmund for the first time after his return to Narbonne: she again takes her son for the "phantom" of his dead father, and pages of confusion culminate in her impassioned interrogation:

> Art thou my husband wing'd from other orbs
> To taunt my soul? What is this dubious form,
> Impress'd with ev'ry feature I adore,
> And every lineament I dread to look on!
> Art thou my dead or living son? (66)

Edmund cannot be resolved into heir, corpse, or apparition: fundamentally "dubious," he is all at once. *The Mysterious Mother* thus intensifies the effect of mingling forms that organizes *The Castle of Otranto*. While Theodore was frequently indistinguishable from both his illegitimate double, Conrad, and his legitimating ancestor, Alfonso, Edmund's indistinguishability from his father is both the sign of his legitimacy and the source of its perverse destruction. This contradiction is captured in the image of the illuminated lamp: while Edmund uses the image to describe the time of inheritance, an endless succession in which a life, extinguished, passes on its flame to be borne by the next generation, the Countess over and over again sees Edmund's dead father "relumed" in the body of her son. This too-literal version of inheritance's structuring mechanisms reveals the normative operations of straight time's "self-naturalizing temporality" as a claustrophobic horror, demonstrating Stephanie Insley Hershinow's observation that "the incest plot scandalizes the marriage plot's heteronormative logic, revealing through the exaggeration of heteronormativity the essential perversity of those norms as such."[72] *The*

Mysterious Mother's conclusion likewise scandalizes Enlightenment. Though Florian believes the clarifying light of rational modernity "shall do mankind and Edmund justice" (101), the hoped-for revelations are monstrous. Rather than projecting a future, here Enlightenment's linked visual and genealogical forms perform instead an endless, obscene recursion. *The Mysterious Mother* thus evokes normative material, social, and narrative forms of futurity only to halt them, deploying the incest plot's ability to figure plot "at its most static and, therefore, at its most resistant to plot as we've tended to understand it (that is, as the kind of desire that moves us inexorably forward)."[73] The "catastrophe" of the play's conclusion ultimately explodes such futures, leaving us to pick up the pieces, to imagine and make something else.[74]

As I have traced across this chapter, Walpole's gothic aesthetics materialize alternate futures through a formal methodology of queer worldmaking. Yet these gothic forms also reconstruct foundational Enlightenment violence. When narrating the Vere family's decline, for instance—the ruin that makes the family's fragmented remains available for remaking at Strawberry Hill—Walpole couches his mockery in anti-Blackness. He writes to George Montagu that when Lord and Lady Vere visited Strawberry Hill, they "called cousins with [the lantern], and would very readily have invited it to Hanworth," where they then lived: "but her *Portuguese* blood has so *blackened* the true stream, that I can't bring myself to offer so fair a gift to their chapel."[75] This reference picks up a strand of gossip that had featured in Walpole's correspondence with Montagu since the 1740s; in these letters he repeatedly speculates (and makes racist jokes) about the current Lady Vere's Portuguese grandmother.[76] And while, for Walpole, the "blackening" of the Vere line extinguishes its futures, his gothic lantern made "of tin japanned" uses an Orientalized material blackness to enclose its English heraldic fragments.[77] Read alongside Walpole's racist disdain for Lady Vere, this aspect of the lantern's construction suggests how the technology's capacity to produce gloomth is literally structured by the blackness excluded from its anticipatory illumination. Walpole's claim to the Vere line may shatter the British family to make a queer utopia, but this utopia is premised on the white supremacy it reconstitutes. Here too we might consider how Walpole's gothic incest plots, as they twist inheritance and halt narrative's forward motion, also perform the consolidation of whiteness followed to its logical conclusion through their extreme resistance to exogamy. Thus, though Carolyn Dinshaw theorizes amateur medievalism of the kind that organizes Walpole's gothic project as an affective form that can "bring out or enact temporal multiplicities" and "point to or even create completely other kinds of time," it is crucial to remember how often such attachments are shaped by a racial logic: by explicit or implicit fantasies of a white past and a white future.[78] What can we make of *these* inheritances, inseparable from Walpole's future vision? Muñoz articulates a queer utopian hermeneutic that is unfaithful to its sources; as he writes about his own use of Ernst Bloch's work as a primary theoretical framework for *Cruising Utopia*, even though "it has been rumored that Bloch did not hold very progressive opinions on issues of gender and sexuality," and about Herbert Marcuse's intellectual engagement with the Nazi philosopher Martin Heidegger, "To draw from such sources and ultimately make them serve another project, one that the author himself would have quickly denounced, serves as a critical engagement—critique as willful disloyalty to the master."[79] Walpole's own gothic methodology stages an encounter with the past that remains flagrantly unfaithful to its forms and the futures they demand, generating instead an elsewhere. We too need to be disloyal to our sources, willing to smash and remake our eighteenth-century archive and so attempt to illuminate other futures.

NOTES

1. Horace Walpole, *A Description of the Villa of Horace Walpole, Youngest Son of Sir Robert Walpole Earl of Orford, at Strawberry-Hill, Near Twickenham. With an Inventory of the Furniture, Pictures, Curiosities, &c.* (Twickenham, UK: Strawberry Hill Press, 1784), 2.

2. Horace Walpole to Horace Mann, Strawberry Hill, April 27, 1753, in *Horace Walpole's Correspondence*, ed. W. S. Lewis, 48 vols. (New Haven, CT: Yale University Press, 1937–1983), 20:372.

3. Horace Walpole to Horace Mann, Strawberry Hill, June 12, 1753, in Lewis, *Horace Walpole's Correspondence*, 20:381. Walpole wrote of the "origin" of *The Castle of Otranto* that "I waked one morning in the beginning of last June from a dream, of which all I could recover was, that I had thought myself in an ancient castle (a very natural dream for a head filled like mine with Gothic story) and that on the uppermost bannister of a great staircase I saw a gigantic hand in armour." Horace Walpole to William Cole, Strawberry Hill, March 9, 1765, in Lewis, *Horace Walpole's Correspondence*, 1:88.

4. Walpole, *Description of the Villa*, i. On the relationship between text and architecture at Strawberry Hill specifically, see also Ruth Mack, "Paper Castle, Paper Collection: Walpole's Extra-Illustrated Copy of the *Description of . . . Strawberry Hill*," in *Horace Walpole's Strawberry Hill*, ed. Michael Snodin and Cynthia Roman (New Haven, CT: Yale University Press, 2009), 107–115; and Luisa Calè, "Extra-Illustration and Ephemera: Altered Books and the Alternative Forms of the Fugitive Page," *Eighteenth-Century Life* 44, no. 2 (2020): 111–135.

5. Jack Halberstam, *In a Queer Time and Place: Transgender Bodies, Subcultural Lives* (New York: New York University Press, 2005), 4. Michael Nicholson has recently analyzed Walpole's collection of his early writing, *Fugitive Pieces in Verse and Prose* (1758), using the framework of queer temporality: see Nicholson, "Fugitive Pieces: Walpole, Byron, and Queer Time," *The Eighteenth Century* 60, no. 2 (2019): 139–162.

6. José Esteban Muñoz, *Cruising Utopia: The Then and There of Queer Futurity* (New York: New York University Press, 2009), 25.

7. Horace Walpole to George Montagu, Strawberry Hill, June 11, 1753, in Lewis, *Horace Walpole's Correspondence*, 9:150–151.

8. Anna Eavis and Michael Peover, *Horace Walpole's Painted Glass at Strawberry Hill* (London: British Society of Master Glass Painters, 1995), 1.

9. Michael Peover, "Stained and Painted Glass," in Snodin and Roman, *Horace Walpole's Strawberry Hill*, 64–65.

10. Richard Marks, "The Reception and Display of Northern European Roundels in England," *Gesta* 37, no. 2 (1998): 220, 218, 219.

11. Horace Walpole to George Montagu, Mistley, July 25, 1748, in Lewis, *Horace Walpole's Correspondence*, 9:67.

12. Thomas Babington Macaulay, *The History of England from the Accession of James II*, vol. 2 (London: Longman, Brown, Green, and Longmans, 1849), 320.

13. Lewis Majendie, *An Account of Hedingham Castle, in the Country of Essex* (London John Nichols, 1796), 1.

14. Elizabeth Freeman, *Time Binds: Queer Temporalities, Queer Histories* (Durham, NC: Duke University Press, 2010), 4.

15. Lee Edelman, *No Future: Queer Theory and the Death Drive* (Durham, NC: Duke University Press, 2004).

16. Halberstam, *In a Queer Time and Place*, 4.

17. Walpole to Montagu, Mistley, July 25, 1748, 9:63.

18. Horace Walpole, *The Castle of Otranto: A Gothic Story*, ed. Nick Groom (Oxford: Oxford University Press, 2014), 5–6. Further references to this edition appear parenthetically in the text.

19. *OED Online*, s.v. "exploded, adj." (Oxford: Oxford University Press), https://www.oed.com.

20. Walter Scott, introduction to *The Castle of Otranto* (1811), by Horace Walpole, in *Horace Walpole: The Critical Heritage*, ed. Peter Sabor (New York: Routledge, 2007), 93. For two different views on this question, see Laura Doyle, "At Liberty's Limits: Walpole and Lewis," in *Freedom's Empire: Race and the Rise of the Novel in Atlantic Modernity, 1640–1940* (Durham, NC: Duke University Press, 2008), 215–229; and Abbey Coykendall, "Chance Enlightenments, Choice Superstitions: Walpole's Historic Doubts and Enlightenment Historicism," *The Eighteenth Century* 54, no. 1 (2013): 53–70.

21. Jane Austen, *Northanger Abbey*, in *Northanger Abbey and Other Stories*, ed. James Kinsley, John Davie, and Claudia L. Johnson (Oxford: Oxford University Press, 2003), 145.

22. *OED Online*, s.vv. "explode, v.," "exploded, adj." (Oxford: Oxford University Press), https://www.oed.com.

23. See George Haggerty, *Horace Walpole's Letters: Masculinity and Friendship in the Eighteenth Century* (Lewisburg, PA: Bucknell University Press, 2011); and Jill Campbell, "'I Am No Giant': Horace Walpole, Heterosexual Incest, and Love among Men," *The Eighteenth Century* 39, no. 3 (1998): 238–260.

24. Haggerty, *Horace Walpole's Letters*, 72.

25. Eugenia Zuroski, "'That You May Show Us What We Have Seen': Bentley's Drawings and the Archive of Queer Feeling," paper presented at "Text Artefact Identity: Horace Walpole and the Queer Eighteenth Century," Twickenham, UK, February 2019.

26. Michael Peover and Kevin Rogers, "New Light on Strawberry Hill: Walpole's Display of Glass and the Representation of the Ancient Abbey," *Journal of Stained Glass* 34 (2010): 10.

27. Horace Walpole to Bentley, Tunbridge, August 5, 1752, in Lewis, *Horace Walpole's Correspondence*, 35:143; Zuroski, "'That You May Show Us What We Have Seen.'"

28. Horace Walpole to Horace Mann, Arlington Street, January 10, 1750 OS, in Lewis, *Horace Walpole's Correspondence*, 20:111.

29. Walpole to Mann, Strawberry Hill, June 12, 1753, 20:381.

30. Andreas Petzold, "Stained Glass in the Age of Neoclassicism: The Case of Eglington Margaret Pearson," *British Art Journal* 2, no. 1 (2000): 58.

31. Alice Tweedy McGrath, "Unaccountable Form: Queer Failure and Jane Barker's Patchwork Method," *The Eighteenth Century* 60, no. 4 (2019): 354.

32. Ibid., 357.

33. Macaulay, *The History of England*, 320.

34. Anne F. Harris, "Glazing and Glossing: Stained Glass as Literary Interpretation," *Journal of Glass Studies* 56 (2014): 303.

35. Horace Walpole to Horace Mann, Strawberry Hill, September 30, 1784, in Lewis, *Horace Walpole's Correspondence*, 25:532.

36. Matthew M. Reeve, "Dickie Bateman and the Gothicization of Old Windsor: Architecture and Sexuality in the Circle of Horace Walpole," *Architectural History* 56 (2013): 115.

37. Ibid., 115–116.

38. Horace Walpole to William Cole, Strawberry Hill, July 12, 1778, in Lewis, *Horace Walpole's Correspondence*, 2:100.

39. Reeve, "Dickie Bateman," 115.

40. Walpole to Mann, Strawberry Hill, April 27, 1753, in Lewis, *Horace Walpole's Correspondence*, 20:372.

41. Muñoz, *Cruising Utopia*, 25.

42. Ibid., 1.

43. Ibid., 28.

44. Ibid., 4.

45. See Eve Kosofsky Sedgwick, *The Coherence of Gothic Conventions* (New York: Methuen, 1986).

46. Eve Kosofsky Sedgwick, "Paranoid Reading and Reparative Reading, or, You're So Paranoid You Probably Think This Essay Is about You," in *Touching Feeling: Affect, Pedagogy, Performativity* (Durham, NC: Duke University Press, 2003), 149.

47. Ibid., 147.

48. Harris, "Glazing and Glossing," 303.

49. *Critical Review*, January 1765, in Sabor, *Horace Walpole*, 69.

50. Walpole to Cole, Strawberry Hill, March 9, 1765, 1:88.

51. Matthew M. Reeve, *Gothic Architecture and Sexuality in the Circle of Horace Walpole* (University Park: Pennsylvania State University Press, 2020), 116.

52. Jason Farr, *Novel Bodies: Disability and Sexuality in Eighteenth-Century British Literature* (Lewisburg, PA: Bucknell University Press, 2019), 18–19.

53. Ibid., 19.

54. Harris, "Glazing and Glossing," 303.

55. Muñoz, *Cruising Utopia*, 1.

56. Sean Silver, "Visiting Strawberry Hill: Horace Walpole's Gothic Historiography," *Eighteenth Century Fiction* 21, no. 4 (2009): 542.

57. Ibid., 556.

58. Ibid., 550.

59. Walpole, *A Description of the Villa*, 2.

60. Muñoz, *Cruising Utopia*, 1.

61. Ibid., 22; Muñoz writes that "A posterior glance at different moments, objects, and spaces might offer us an anticipatory illumination of queerness."

62. Francis Bacon, *The New Organon*, trans. and ed. Lisa Jardine and Michael Silverthorne (Cambridge: Cambridge University Press, 2000), 100.

63. Robert Hooke, *Micrographia; or, Some Physiological Descriptions of Minute Bodies Made by Magnifying Glasses* (London: Martyn and Allestree, 1665), v, viii–ix.

64. A.D.C. Simpson, "Reeve, Richard (d. 1666), Maker of Optical Instruments," in *Oxford Dictionary of National Biography: In Association with the British Academy: From the Earliest Times to the Year 2000*, edited by H. C. G. Matthew and Brian Harrison (Oxford: Oxford University Press, 2004), https://doi.org/10.1093/ref:odnb/48372.

65. Samuel Pepys, "19 August 1666," in *The Diary of Samuel Pepys: A Selection*, ed. Robert Latham (London: Penguin, 2003), 655.

66. Thomas L. Hankins and Robert J. Silverman, *Instruments and the Imagination* (Princeton, NJ: Princeton University Press, 1995), 70.

67. Koen Vermeir, "The Magic of the Magic Lantern (1660–1700): On Analogical Demonstration and the Visualization of the Invisible," *British Society for the History of Science* 38, no. 2 (2005): 135.

68. Terry Castle, "Phantasmagoria: Spectral Technology and the Metaphorics of Modern Reverie," *Critical Inquiry* 15, no. 1 (1988): 27. For a sustained analysis of the relationship between gothic fiction and optical technologies like the magic lantern, see David J. Jones, *Gothic Machine: Textualities, Pre-cinematic Media and Film in Popular Visual Culture, 1670–1910* (Cardiff: University of Wales Press, 2011).

69. Jill H. Casid, *Scenes of Projection: Recasting the Enlightenment Subject* (Minneapolis: University of Minnesota Press, 2014), 3.

70. Muñoz, *Cruising Utopia*, 25.

71. Horace Walpole, *The Mysterious Mother: A Tragedy* (Twickenham, UK: Strawberry Hill Press, 1768), 100–101. Further references to this edition appear parenthetically in the text.

72. Stephanie Insley Hershinow, "The Incest Plot: Marriage, Closure, and the Novel's Endogamy," *The Eighteenth Century* 61, no. 2 (2020): 154.

73. Ibid., 150.

74. Walpole writes in his "Postscript" to the play that "Every scene tends to bring on the catastrophe, and the story is never interrupted or diverted from its course" as part of his

assertion that the work follows the dramatic "rules laid down by the critics." Horace Walpole, "Postscript," in *The Mysterious Mother*, 7–8.

75. Walpole to Montagu, Strawberry Hill, June 11, 1753, in Lewis, *Horace Walpole's Correspondence*, 9:151.

76. See Horace Walpole to George Montagu, Mistley, July 14, 1748, in Lewis, *Horace Walpole's Correspondence*, 9:59–62; Walpole to Montagu, Mistley, July 25, 1748, ibid., 9:63–67; and Walpole to Montagu, Strawberry Hill, August 11, 1748, ibid., 9:67–73.

77. For a discussion of the conceptual and material links between Orientalism and anti-Blackness in the long eighteenth century, see Chi-ming Yang, "Asia Out of Place: The Aesthetics of Incorruptibility in Behn's *Oroonoko*," *Eighteenth-Century Studies* 42, no. 2 (2009): 235–253.

78. Carolyn Dinshaw, *How Soon Is Now? Medieval Texts, Amateur Readers, and the Queerness of Time* (Durham, NC: Duke University Press, 2012), 29. See also Helen Young, "Race, Medievalism and the Eighteenth-Century Gothic Turn," *Postmedieval: A Journal of Medieval Cultural Studies* 11, no. 4 (2020): 468–475; and recent reckonings with the structuring racism of medieval studies in, for example, Dorothy Kim, "Introduction to Literature Compass Special Cluster: Critical Race and the Middle Ages," *Literature Compass* 16, nos. 9–10 (2019): 1–16.

79. Muñoz, *Cruising Utopia*, 2, 17.

CHAPTER 7

Telegraphic Supremacy in Maria Edgeworth's "Lame Jervas"

Deven M. Parker

In 2018, the U.S. Congress passed the National Defense Authorization Act to prohibit government agencies from contracting with the Chinese telecom giant Huawei, the largest smartphone company in the world. The ban was passed on the grounds that Huawei could potentially use its devices to spy on Americans for the Chinese state.[1] Despite Huawei's persistent denials of state collusion, Congress instituted the technology ban "for the purposes of public safety, security of government facilities, physical security surveillance of critical infrastructure, and other national security purposes."[2] It also urged American consumers and businesses not to purchase Huawei's phones despite the brand's ubiquity in global markets. Of course, this was not the first time that national security concerns prompted suspicion of foreign technologies, nor the first time that some people feared that communication technologies would also facilitate dangerous forms of surveillance. Indeed, only six years prior to the Huawei ban, the United States itself was implicated in cell-phone spying.[3] Even more recently, the Israeli surveillance company NSO Group came under fire for selling its hacking spyware, Pegasus, to authoritarian governments across the globe.[4]

Situating these events within a broad historical context, this essay explores how anxieties about foreign communication technologies were as prevalent in eighteenth-century Britain as they are now—and what this period's literature can reveal to us about emerging technologies and the national interests they serve. "Lame Jervas," the novella occupying the first volume of Maria Edgeworth's *Popular Tales* (1804), serves as a case study in this essay for exploring early Romantic attempts to neutralize the disruptive potential of a particular technology: the optical telegraph. Building on recent work by Mary Fairclough, Aileen Douglas, and others who have analyzed representations of this unique technology, I situate "Lame Jervas" in a period of intense preoccupation in newspapers and other media with the optical telegraph, which was perceived in the final decade of the eighteenth century as a threat both close to home and throughout the empire.[5] British writers and cartoonists feared that it might further the agendas not just of Jacobin radicals but also of Revolutionary France's key military ally in India, Tipu Sultan. Responding to these concerns, Edgeworth recasts the telegraph in her novella as a signifier of Britain's imperial stability and technological superiority at a time when neither was certain.

Edgeworth's titular Jervas, an English inventor and man of science, travels to India to exhibit inventions to "Tippoo Sultan," an Orientalized despot based on the real ruler who, in the 1790s, controlled vast swaths of India and fought a series of wars with the East India Company.[6] Edgeworth's fictional Tippoo eagerly embraces the technologies presented to him by the English narrator and is especially enthusiastic about Jervas's portable telegraph. Edgeworth's father, the Anglo-Irish statesman Richard Lovell Edgeworth, had invented a version of this device years earlier, but it was never officially adopted by the state. As I show, his efforts were part of a broader campaign in the 1790s to establish a telegraph network in England and Ireland as a precaution against invasion while Britain was at war with France. France had recently established its own telegraph network, which quickly became a symbol of Jacobinism and a sign, thanks to the British press, of France's military gains over England. Facing pressure to compete with France, the administration of Prime Minister William Pitt began soliciting designs for a telegraph that was as effective as France's but distinctly British in its invention and design. Lovell Edgeworth, with his daughter's assistance, published multiple pamphlets advocating for the state's adoption of his telegraph.[7]

"Lame Jervas" channels national anxiety about foreign technological advances through, paradoxically, its celebration of British technological ingenuity. By Edgeworth's having Tippoo Sultan take up her father's telegraph, she casts the technology as a distinctly *British* export, joining a chorus of writers, including her father, who, in the face of fears about perceived French military gains thanks to the telegraph, endeavored to show that the device was in fact a local invention rather than a dangerous foreign import. In the fight to re-imagine the telegraph's origins, "Lame Jervas" portrays the telegraph and other technologies as a means of expanding Britain's imperial influence at a time when the nation was seen as falling behind others, particularly France, in a race for technological supremacy.

Given the optical telegraph's relative obscurity, especially compared to the nineteenth-century device of the same name, it is worth recounting its origins. Using either windmill-like arms (the French version) or shutters (the British version), the movements of optical telegraphs represented coded words or letters. Telegraph operators used telescopes to record the movements of a telegraph situated several miles away and then repeated the message to the next station in the network until it arrived at its destination. The telegraph network was installed in England in 1795 in order to transmit information between the southern coast and London about enemy movements in the Channel. Unfortunately for operators, foggy weather on the English coast often interfered with its use.[8] The network was a public medium for private information as it visibly conveyed coded messages. Its messages were hidden in plain sight, visible to everyone but intelligible only to a privileged minority of officers. Those who observed the movements of the device set atop the office of the Admiralty at Charing Cross would have known that a secret message was being conveyed before their eyes.

Joseph Lakanal, one of the leaders of the French Revolution, famously tested France's telegraph on July 12, 1793, sending a message from Saint Martin-du-Tertre to Ménilmontant—a distance of over fifty-one kilometers—in just nine minutes.[9] Advocating for the adoption and installation of the machine, which was first invented by Claude Chappe, Lakanal later wrote, "what brilliant destiny do science and the arts not reserve for a republic which, by the genius of its inhabitants, is called to become the nation to instruct Europe."[10] His prediction came true: not only did France's revolutionary fervor spread across the world, but by the middle of the nineteenth century, most nations in Europe had erected telegraph systems of their own, to varying degrees of success. Yet none matched the efficiency of the French system: messages sent from Paris reached the nation's borders in about three to four hours, compared to the several days it took messengers on horseback. The speed at which French messages could be conveyed shocked the international community when, in September 1794, papers reported that the National Convention received the telegraphic message that it had captured the Royalist outpost of Condé—263 kilometers from Paris—on the same day it happened.[11]

Britain's development and implementation of its own optical telegraph system was motivated by British newspapers, whose coverage of the French telegraph's effectiveness culminated in a media frenzy that ultimately influenced the Pitt administration's adoption of the device and even the design of the technology itself. The first mention (that I can find) of the French telegraph in English newspapers appeared on September 18, 1794, in the *British Evening Post*; it conveys significant anxiety about the foreign invention's speed and efficiency: "If we are to believe the Dutch accounts, the French Telegraph is much superior to what it has hitherto been conceived to be. They say, that an account of the surrender of Sluys was sent to Paris, and an answer from the Committee of Publick Safety received by the Representatives of the People with the army, in *half an hour* after the capitulation."[12] The report appears directly after a notice about continued mob brutality in Paris, where France was at the height of its notorious Reign of Terror. Read within this context, the report on the telegraph is presented as evidence of the dangerous threat that France represented to English national security. The paper cites the source of its intelligence as foreign, a kind of international hearsay that leaves room for speculation and possible disbelief on the part of the reader. Yet even as the report's secondhand nature seems designed to undercut its authenticity, the italicized "*half an hour*" evokes fear and awe of what the telegraph may be capable of, sensationalizing its capacity for speed.

In the months that followed this initial notice, British papers regularly reported on France's effective use of telegraphy, continuing to marvel at its speed and linking these reports with the enemy's continued victories. One notice from the *Courier and Evening Gazette* reads that "the intelligence of the surrender of Saas-de-Grand, Phillipine, Hulst, and Axel was reported to the Convention in their sitting of the 25th by the telegraph, which conveyed the intelligence in 50 minutes from Lisle to Paris," while days later the *Morning Chronicle* related that "the Telegraph continues to announce victories obtained or fortresses taken."[13] These notices conflate France's rapid series of military victories with the telegraph's power to convey this news back to Paris. Even though the telegraph merely reported on these victories—just as the papers themselves did—the British press suggested a causal link between France's military successes and its telegraph network. Interestingly, British newspapers reported on the phenomenon of reporting itself, implying that communication networks were an integral part of French military might and as newsworthy as any battle. The French telegraph thus came to stand for more than the device itself: it embodied national anxiety about France's military gains, a possible French invasion of England, and, from an ideological perspective, the power of Jacobin idealism.

Thanks to this flurry of reports in autumn of 1794, the British public understandably yearned to see and experience the marvels of the French telegraph, even as it worried that the rival nation had a leg up on military technology. The papers fed this desire by offering detailed descriptions of the device, as in a November 26 article in the *Courier and Evening Gazette* that described how "a proof has been made of the telegraph at the Palace . . . to convey intelligence through Trattberg to Drottningholm at the distance of one league and an half, in a few moments. The machine consists in a high piece of timber, with two moveable cross beams on each side; by the nature of their movements the intelligence is understood."[14] In addition to descriptions of the telegraph's application and appearance, others offered the chance to view the inimitable invention in person. The *Morning Post*, for example, advertised a telegraph exhibition at Astley's New Circus, claiming that "Mr. Astley . . . flatters himself that he is perfectly acquainted with the nature and construction of the TELEGRAPHE, so much the conversation, astonishment, and admiration of the present day. [He] will exhibit it for a few evenings, at the Lyceum, in the Strand."[15] Advertisements like this also speak to the climate of secrecy that surrounded the French telegraph, contributing to its hype. Even though it was widely reported on, few knew what it looked like, and almost none had seen it in person. Those like Mr. Astley, who claimed familiarity with its appearance and workings, could be seen to possess privileged intelligence, a window into the secret source of France's military prowess.

Coupled with the public's eagerness for news about the telegraph and glimpses of its appearance was the conflicting sense that this technology, as a physical manifestation of radicalism and foreign violence, was at once something to fear but also something that should be replicated in England. We get a glimpse of this sentiment in James Gillray's print of January 26, 1795, "French-Telegraph Making Signals in the Dark" (figure 7.1). The print shows William Pitt's Whig rival, Charles James Fox, as a human French telegraph helping the enemy fleet invade England, to critique Fox's support for the Revolution. Fox's limbs become the characteristic windmill arms of the telegraph, which he uses to shine light on the French fleet

Figure 7.1. James Gillray, "French-Telegraph Making Signals in the Dark." Hand-colored etching and aquatint. London: Published for H. Humphrey, 1795. B1981.25.907. Courtesy of the Yale Center for British Art, Paul Mellon Collection.

and provide navigational assistance to the invading army. Gillray depicts the French telegraph as a military weapon and, crucially, an extension of the nation's new violent regime that threatens to infiltrate Britain both physically and ideologically. By supposedly embracing Jacobin radicalism, Fox physically transforms into its most powerful weapon and symbol.

As telegraph depictions proliferated between 1794 and 1795, Britain saw a wave of support for adopting the telegraph in the country, despite the device's association with Jacobinism. Many Britons felt that their nation needed to create a telegraph of its own to gain military ground over France; however, this required a delicate balance of replicating the French invention while making it distinctly British in order to evacuate the technology of the dangerous ideology that, thanks to Gillray and others, it had come to embody. As Mary Fairclough argues in her analysis of telegraph representations from this period, the "strong impulse to adapt the telegraph for use in Britain demonstrates the importance of removing the stigma of appropriating a French revolutionary machine."[16] The prolific theater entertainer Charles Dibdin called for precisely this balance of technological borrowing and remaking in a song about the telegraph that appeared in his 1794 *Great News, or a Trip to the Antipodes*:

If you'll only just promise you'll none of you laugh,
I'll be after explaining the French telegraph!
A machine that's endow'd with such wonderful pow'r,
It writes, reads, and sends news fifty miles in an hour.
Then there's watch-words, a spy-glass, and index on hand,
And many things more none of us understand;
But which, like the nose on your face, will be clear,
When we have, as usual, improv'd on them here.
. . . .
Adieu, penny-posts! mails and coaches, adieu;
Your occupations' gone, 'tis all over wid you:
In your place, telegraphs on our houses we'll see,
To tell time, conduct lightning, dry shirts, and send news.
Then, while signals and flags stream on top of each street,
The town to a bird will appear a grand fleet:
And since England's grand fleet to the French conveys fear,
Sure, shan't we improve on the telegraph here?[17]

The initial lines of Dibdin's song echo the awe and fear of those early news reports, speculating about the telegraph's unprecedented powers and speed and claiming to have special access to its secret workings. Yet, claims the song, the telegraph's special abilities will only become clear to the British public once the government decides to implement and "improve on" the technology, making it British and thereby rendering it secure. With tongue-in-cheek wit, Dibdin imagines a London in which telegraphs replace "penny-posts," "mails," and "coaches," occupying rooftops across the city. The cumulative effect of introducing telegraphs in London, he argues, will be to reverse the fear with which the English currently regard the French telegraph. Rather than the mysterious foreign technology inciting fear, the appearance of London's telegraphs will look like a "grand fleet" of ships that will strike fear into the French and win the war. Making a case for a British reworking of the French telegraph instead of inciting fear of the technology as in Gillray's depiction, Dibdin concludes his song by putting pressure on the government to "improve on the telegraph" as a show of military might and patriotic pride.

Implicit within the humorous song, however, is the concern that Britain will not be able to keep pace with France's technological ingenuity and will suffer military losses as a result. The ideas in Dibdin's song were echoed by a series of articles in *The Gentleman's Magazine*, which published reader suggestions on how the French machine could be introduced and remade in a distinctly "English" visage.[18] In the face of such pressure, the British Admiralty introduced its own telegraph network in 1795. Designed by Lord George Murray, it purportedly worked just like the French system but looked different enough that it could not be associated with French politics. Rather than arms, Murray's telegraph deployed a system of shutters that flipped between horizontal and vertical positions to create unique signals, a method that was unfortunately more difficult to decipher at a distance.

Even after Britain's first telegraph line was erected between London and Deal the following year, many Britons continued to grapple with the embarrassing fact that France had beat them to the punch and that Britain continued to lag technologically, with only one line in its telegraph network compared to the many that France had established throughout Europe. Indeed, many betrayed this apparent insecurity by attempting to prove that England had first invented the idea for the technology, even if it had not been first to build or implement it. Newspapers presented such historical revisionism under the guise of comparing different telegraph systems, as when the *Whitehall Evening Post* reprinted a lengthy excerpt of the scientist Robert Hooke's 1684 discourse to the Royal Society in which the inventor proposed a plan for "shewing a way how to communicate one's mind at a great distance."[19] Hooke's treatise was republished in the *Post* as telegraph mania swept the nation; the paper casts it as the intellectual predecessor of the French invention, suggesting to readers that a British inventor had at least first thought of the idea, even if he never built it. On November 29, 1794, the *Morning Post* reported that a "telegraphic experiment" had recently taken place in Ireland by an inventor who, like Hooke, had supposedly beaten the French in the race for rapid communications. The report insists that the technology "was originally invented by Mr. Edgeworth, of the county of Longford, though first put into real use by the French."[20] As with the report on Hooke, this report insists on the separation—and the supremacy—of the *idea* of the telegraph from its practical application; while France may have beaten Britain to the punch with the latter, it is the primacy of ideas that supposedly matters and, not so subtly, provides evidence of Britain's superiority.

Indeed, Richard Lovell Edgeworth and Maria Edgeworth claimed in multiple essays and memoirs that Lovell Edgeworth had first invented the telegraph. In contrast to the French design, Lovell Edgeworth's portable version employed a system of triangle pointers that indicated combinations of coded symbols. The inventor sought to have the device installed in Ireland, where, during a series of meetings with Irish leaders in 1795 and 1796, he argued that it offered a solution to the pressing concern that the French would attempt an invasion of Ireland's shores, where, unlike England, they might find a populace sympathetic to their radical ideals.[21] Much to the Edgeworths' disappointment, the proposal was rejected. In 1797, Lovell Edgeworth collaborated with his daughter Maria and published a letter to the Earl of Charlemont complaining about the rejection and arguing for the superiority of his invention's design over others'.[22] This publication, along with an essay on the subject that appeared the same year in the *Transactions of the Royal Irish Academy*, claims the invention as Lovell Edgeworth's intellectual property and attempts to rewrite its origins in order to claim it for Britain rather than France.[23] Together, the Edgeworths' essays on telegraphy echo many of the concerns about the dangerous influence of foreign technology expressed in the London newspapers; they also provide an immediate context for Maria Edgeworth's depiction of telegraphy and technology more generally in "Lame Jervas."

In the *Letter to the Right Hon. The Earl of Charlemont, on the Tellograph, and on the Defence of Ireland*, the origin story that the Edgeworths assign to the telegraph is crafted to show that Lovell Edgeworth's invention preceded Claude Chappe's, as well as Murray's, Gamble's, or that of any other inventor who staked a claim to it. The Edgeworths offer the following anecdote:

> When I heard of the French Telegraph a new object arose for my exertions. I recalled to my mind experiments that I had tried so long ago as the year 1767, when I had practiced this species of aerial communication; and thinking that it might be peculiarly useful to this country, I constructed some machines with which I conversed, in August 1794, between Packenham Hall, the seat of Lord Longford, and Edgeworthstown. Finding my success equal to my expectations, I was advised by the present Bishop of Ossory . . . to shew my invention to some gentleman, whose opinion would be attended to by administration.[24]

Even though the purpose of the essay in question is supposedly to demonstrate that Lovell Edgeworth had been misled and poorly treated by Irish authorities in having his telegraph design rejected, they foreground their argument with evidence of Lovell Edgeworth's status as the first inventor of the device, tracing the invention back to 1767. Lovell Edgeworth would double down on this account in his *Memoirs*, explaining how his telegraph was originally conceived as part of a bet.[25] He claims, however, that although he had materially invented the technology, he did not realize how useful it could potentially be until 1794, when France installed its first network. This narrative effectively undercuts the primacy of the French telegraph even while acknowledging it; in other words, it claims that the *idea*, which is supposedly more important than the thing itself, belongs to him.

In a separate essay that appeared in *Transactions of the Royal Irish Academy*—also penned with Maria Edgeworth—the inventor not only reasserts the originality of his telegraph design but positions it as part of a longer genealogy of inventions and ideas relating to communicating at a distance. This move serves to legitimize and authenticate his invention and to counter the idea that it is a foreign import. "The art of conveying intelligence by sounds and signals is of the highest antiquity," they write. "It was practiced by Theseus in the Argonautic expedition, by Agamemnon at the siege of Troy, and by Mardonius in the time of Xerxes. . . . The signals of Agamemnon are beautifully described in Aeschylus, and every station is pointed out with geographical accuracy. Tamerlane's telegraphy was not very refined, but it was sufficiently intelligible."[26] The two go on to argue that even the oracles at Delphi should be regarded as a kind of early telegraphic system, through which priests "pretended some secret mode of receiving intelligence."[27] The essay positions Lovell Edgeworth as the intellectual inheritor and successor of these classical military heroes and legendary figures, the modern perfecter of their "beautiful" yet primitive attempts to communicate at a distance. Placing himself within this noble genealogy of telegraphic inventors offers a respectable way of bypassing and downplaying the French inventor Chappe's supposed ingenuity. When the essay arrives at modern France in describing the historical trajectory of telegraphy, the description of Chappe's device downplays its effectiveness: "It is sufficient to say that it consists of a tall pole, with three movable arms, which may be seen at a considerable distance through telescopes; . . . It is obviously liable to mistakes, from the number of changes requisite for each word, and from the velocity with which it must be moved to convey intelligence with any tolerable expedition."[28] The simplistic description of the French telegraph contrasts with the sensationalized depictions that appeared in the English papers, and the Edgeworths emphasize the machine's fallibility over the reality that it was the most effective long-distance signaling network ever developed, especially compared to Britain's error-prone one. In attempting to wrest telegraphy from foreign hands, the authors craft an alternative history in which the French are merely adopters rather than inventors. At the same time, Lovell Edgeworth responds to his telegraph's critics and his rejection at the hands of the Irish government, claiming that "the disposition to ridicule every scientific project as absurd until it has been absolutely brought to perfection has been the common topic of complaint amongst men of inventive genius," a category in which Lovell Edgeworth includes himself.[29]

In addition to historicizing the telegraph within a classical framework to mask its associations with radicalism and otherness, the Edgeworths' Royal Academy essay posits the invention as an integral tool in Britain's imperial project. As a symbol of classical culture perfected by modern inventors like Lovell Edgeworth, the telegraph is a force for spreading British power across the world. Telegraphy could achieve this, the writers claim, by serving as a "universal language"—it could erase linguistic differences between nations and promote easier communication and, by extension, commerce between Britain and the East. Rather than learn a new language, a man "who wishes to convey his ideas to foreigners who do not understand his language, or whose language he does not understand" could use telegraphy's nonverbal symbols to express himself.[30] Notably, the essay does not advance the telegraph as a means of discourse or dialogue between people of different languages but as a one-way medium for "conveying" the ideas of a man who is undoubtedly European "to foreigners." This unidirectional model of "universal" communication could help spread Western ideas and progress across the globe and, more locally, aid Britain in its military and economic domination of India: "I scarcely dare to foretell," the essay continues, "that a communication by Telegraphs between Europe and the East Indies will one day be established, and that its effects will be more beneficial to Europe than monopoly or conquest."[31]

Reading Edgeworth's "Lame Jervas" in the context of these collaborative essays about the telegraph—which formed part of a larger national effort to recast the technology as ideologically and culturally British rather than French—the novella's celebration of its titular inventor's ingenuity works to reassert Britain's supremacy in terms of both technological prowess and morality, linking the two in the process. From the beginning of the tale, interest in technology and talent with machines indicate upright character and moral virtue. Indeed, when we encounter Mr. Y, one of Jervas's many benevolent employers, technology serves as the focal point in a scene of domestic affection: "I found the good old gentleman and some of his friends in his study, with his children about him; one little chap on his knee, another climbing on the arm of his chair, and two bigger lads were busy looking at a glass tube, which he was shewing them when I came in."[32]

In the context of this tableau, the glass tube—later revealed to be a thermometer—appears not just as a scientific tool but as a sign of filial care. Mr. Y is revealed to be a virtuous man who values his machines and inventions not simply because they increase his profits and improve workers' productivity but also because they elevate the mind and foster fellow feeling. When the uneducated Jervas enters his study, he says, "[Mr. Y] saw that I gazed, with vast curiosity, at several objects in the room, which were new to me."[33] Showing him the thermometer, Jervas continues, "Mr. Y—took great pains to shew me how, and on what occasions, this instrument might be useful."[34] The gentleman's patience with Jervas and willingness to exhibit his devices teach the younger man to trust him and, eventually, to model himself after him. Taking Mr. Y's advice to "make himself useful," Jervas decides to "make a sort of model of the tin-mine" for the amusement of his master's children.[35] After Jervas exhibits his mechanical mine to the family, Mr. Y praises him and encourages further technological pursuits in the service of Jervas's self-improvement, gifting the young man with a magnifying glass.

Beginning with this invention, Jervas's technological skill is what allows him to rise through society's ranks, eventually becoming a gentleman despite his low status at birth. Mr. Y secures him a job as an assistant to a lecturer who travels the country exhibiting "models of the machines used in manufactories" for the amusement and education of gentlemen.[36] Initially, the lecturer treats Jervas with disdain because of Jervas's lack of formal education and low birth; however, this changes when, once again, Jervas's knack for technology and science endears him to his new employer. Recalling an experiment with metal that his master had conducted, Jervas discovers a guinea covered in quicksilver that the man had mistaken for a shilling, causing the lecturer to show him kindness.

Jervas's mechanical exhibitions eventually lead him to meet a director of the East India Company. In the manner of his other employer, the man recognizes both Jervas's technological talent and moral integrity—which, again, go hand in hand—and offers him a place teaching at Dr. Bell's school for orphans in Madras. Before departing, Jervas learns that his first employer, Mr. R, has sent him a farewell gift in the form of various scientific inventions to bring on his journey. These include "two small globes, siphons, prisms, an air-gun and an air-pump, a speaking trumpet, a small apparatus for showing the gases, . . . small balloon, and a portable telegraph, in the form of an umbrella."[37] This catalog of inventions that Jervas brings to India serves as evidence of his employer's kindness, Jervas's talent, and, above all, a reminder of the many inventions that British men of science, like Lovell Edgeworth, had pioneered and popularized during this time. When Jervas arrives in India with the inventions, Dr. Bell is impressed by his knowledge and proposes that Jervas travel as part of a diplomatic mission to the nearby court at Seringapatam to "exhibit some new wonders" for the entertainment of Tippoo Sultan.[38] Serving as an ambassador for his country, Jervas selects the inventions he thinks will most impress and intimidate the foreign ruler: "I packed up my speaking trumpet, my apparatus for freezing water, and that for exhibiting the gasses, my balloons and telegraph, and with these and my model of the tin-mine . . . I set out with two of his eldest scholars upon our expedition."[39]

When Jervas meets Tippoo Sultan, he is disgusted by the ruler's wrath, greed, and arrogance. Like other works of the 1790s set in India, Tipu Sultan's "oriental despotism" serves to justify British imperialism at a time when, argues Kate Teltscher, "challenges to British power were real."[40] Along with France's perceived military gains over Britain that simultaneously occurred on the continent, Teltscher explains how "the sultanate of Mysore, during its period of expansion under Haider Ali and Tipu Sultan, seriously shook British authority in the south. Placed in this context, emphatic assertions of British invulnerability sound more like attempts to bolster national confidence."[41] Stories like Edgeworth's thus promote "confidence boosting images of British rule constructed in the face of such threats," putting forth "anxious assertions of power."[42] Tipu Sultan represented a threat to the East India Company's military and economic power in India thanks in part to his technological prowess and integration of state-of-the-art military devices at a time when, as we have seen, Britain took up such advances at a relatively slow pace compared to other nations. The ruler was a pioneer of artillery, introducing rockets with iron tubes that enabled higher thrust and longer range, a design that vastly outstripped the outdated models used by the East India Company and that would eventually be used by the British themselves in the Napoleonic Wars. Tipu Sultan's promotion of technology also extended to automaton entertainments like the ones Jervas creates. As a show of his power, the ruler—also known as the "Tiger of Mysore"—commissioned a life-size mechanical tiger, which, with help of a spring, mauled a British soldier.

These realities, coupled with the fact that Tipu Sultan was a close ally of France—"one of the aspects of Tipu's power which most alarmed the British," writes Teltscher—illuminate the insecurity that pervades Edgeworth's portrayal of British technological prowess.[43] When Jervas exhibits his inventions to the court,

> The apparatus for freezing, which we then exhibited, seemed to please [Tippoo Sultan]; but I observed that he was, during a great part of the time whilst I was explaining it, intent upon something else; and no sooner had I done speaking than he caused to be produced the condensing engines, made by himself, which . . . he said spouted water higher than any of ours. The sultan, I perceived, was much more intent upon displaying his small stock of mechanical knowledge than upon increasing it; and the mixture of vanity and ignorance, which he displayed upon this and many subsequent occasions, considerably lessened the awe . . . in my mind. Sometimes he would put himself in competition with me, to show his courtiers his superiority; but failing in these attempts, he would then treat me as a species of mechanic juggler, who was fit only to exhibit for the amusement of his court. When he saw my speaking-trumpet, which was made of copper, he at first looked at it with great scorn, and ordered his trumpeters to show me theirs, which were made of silver. . . . But, upon trial, mine was found to be far superior to the sultan's.[44]

Here, Jervas's machines—his freezing apparatus and speaking trumpet—are continually asserted to be more advanced and effective than Tippoo's. While Jervas purports to have come to the palace with pedagogical aims, Tippoo sees his demonstrations for what they really are: performances of cultural superiority. As Jervas displays his technologies, he accuses the sultan of doing the same: rather than passively absorbing Jervas's superior knowledge, Tippoo displays his knowledge in turn. Jervas hastily assures the reader, however, that even though Tippoo possesses similar instruments, the British inventions are "far superior." As the stereotypical Oriental despot—a role in which Edgeworth casts him—Tippoo eventually acknowledges that Jervas's toys are better and, like a child, must have them: "He accepted [the speaking trumpet] with the eagerness of a child who has begged and obtained a new plaything."[45] Yet, even as Tippoo's "vanity and ignorance" disgust Jervas, the English inventor has succeeded in his efforts to spread British technologies and ideas, a crucial aspect of the nation's imperial project. Importing Britishness into India by way of his technologies, Jervas asserts his nation's supremacy to the ruler who, in reality, rendered it most insecure.

The technology that Tippoo desires more than all others is Jervas's portable telegraph. Unlike the freezing apparatus or the speaking trumpet, this device, we learn, is unfamiliar to the despot. In Tippoo's eyes, the telegraph is the crown jewel of technological advancement. Whereas the governments of Britain and Ireland were slow to implement the invention, the fictional Tippoo immediately seizes upon it. Speaking with Prince Abdul Calie, Tippoo's son and one of the princes famously taken hostage by the British years earlier, Jervas explains how the idea to exhibit the telegraph first came to him:

> [The prince] then said to me, "The sultan, my father, is at this time so intent upon preparations for war, that even I should despair of being listened to on any other subject. But you have in your possession, as I recollect, what might be useful to him either in war or peace; and if you desire it, I will speak of this machine to the sultan." . . . He explained to me that he meant my portable telegraph, which would be of infinite use to Tippoo in conveying orders of intelligence across the deserts. . . . A few hours after this conversation, I was summoned into the sultan's presence. His impatience to make trial of the telegraphs was excessive; and I . . . instantly became a person of the greatest importance. The trial of the telegraphs succeeded beyond even my expectations; and the sultan was in a species of ecstasy on the occasion.[46]

Like other depictions of Tippoo, his sons are portrayed as European in their calm, civilized demeanors, perhaps having learned British manners after spending time as hostages of the East India Company.[47] With insight into his despotic father and admiration for Jervas, Abdul Calie suggests that his father might appreciate the telegraph, showing that the prince is civilized because he can appreciate the machine's value beyond its military use, just as Edgeworth and Lovell Edgeworth had argued in their Royal Society essay. Tippoo, too, immediately realizes the ingenuity of Jervas's telegraphs, elevating the inventor to a status of "greatest importance." In light of the real-life Tipu Sultan's alliance with France, this scene reads as a warning about the dangers of telegraphy in the hands of Britain's enemies, particularly when Lovell Edgeworth's own countrymen were reluctant to take up the technology—as Joanna Wharton puts it, Edgeworth warns us that "if the British did not take up such innovations, their enemies would."[48] At the same time, Edgeworth's description of Tippoo's ecstasy upon witnessing the telegraph trial serves to celebrate Jervas's—and Lovell Edgeworth's—distinctly British invention. Even the sultan, who previously tried to downplay the ingenuity of the inventor's instruments, cannot ignore the brilliance of this British machine. While Jervas does not civilize the tyrant, he does succeed in building an alliance with him, where other British emissaries had been unsuccessful. By introducing the telegraph to Tippoo's court, Jervas does not aid the enemy so much as he ensures that any power that Tippoo derives from the telegraph network has been the result of Britain's cultural supremacy. "Lame Jervas's" representation of the telegraph overcompensates for the reality of what Ashley Cohen calls "the global Jacobin crisis," the period in the 1790s when Tipu Sultan, France, and other radical powers across the world threatened British security and national interests.[49] Instead, Edgeworth offers us a version of real-life events in which Britain clearly has the upper hand in both foreign influence and technological advancement, once again rewriting the telegraph's origin story in order to divorce it from Jacobinism. Echoing those earlier news reports that insisted on the British origins of the telegraph—as well as Edgeworth's essays penned with her father—her imagining of the telegraph's introduction in India makes clear that the idea for the invention, if not its practical application, first came from a British mind.

Yet, in the scene that follows, Edgeworth also offers a warning about the dangers of telegraphic technology in the hands of a despot like Tipu—and, by extension, the French. During the initial trial of the telegraphs, Tippoo goes into a rage and nearly kills one of his enslaved men after the man incorrectly transmits a message:

> I had previously instructed Saheb in what he was to do; but, from want of practice, he made some mistake, which threw Tippoo into such a transport of passion, that he instantly ordered the slave's head to be cut off! a sentence which would infallibly have been executed, if I had not represented that it would be expedient to suffer his head to remain on his shoulders till the message was delivered by his telegraph; because there was no one present who could immediately supply his place. Saheb then read off his message without making any new blunder; and the moment the exhibition was over, I threw myself at the feet of the sultan, and implored him to pardon Saheb. I was not likely at this moment to be refused such a *trifle*! Saheb was pardoned.[50]

Sharon Alker has argued that this scene, and the story as a whole, warn of the dangers of technologies like the telegraph falling into the wrong hands: "the story suggests that to prevent the destructive force of dangerous technologies in the hands of 'fanatics' and 'criminals,' technological innovation must be carefully harnessed to diminish its danger."[51] At the same time, however, Edgeworth nullifies this concern about Tippoo's potential to use the telegraph for destruction by suggesting that the ruler's barbarity—his readiness to decapitate Saheb for making a transmission error—might perhaps derail the ruler's successful adoption of a telegraph network. If Tippoo's order had been carried out, he would have executed the man even though "there was no one present who could immediately supply his place" in the network. It is only thanks to Jervas's intervention that the successful trial of the machine takes place, casting doubt on Tippoo's ability to effectively make use of this invention. In this moment, the ruler's bloodlust echoes those newspaper reports of French mob violence—complete with beheadings—that accompanied articles about the Jacobin radicals' successful telegraph installation. As though responding to concerns generated by such reports, here Edgeworth effectively quashes anxiety about telegraphs in the hands of violent leaders by suggesting that these leaders' brutality might hinder their efficient use of the technology. Once again, Edgeworth posits morality and benevolence—embodied by Jervas—as necessary for the effective use of technology.

Edgeworth's depiction of the telegraph echoes the arguments of those writers, including her father, who sought to evacuate the telegraph of radicalism by reimagining its origins; she also furthers this mission by reinscribing the technology itself with Enlightenment values. Tippoo Sultan views the telegraph only as a weapon, no different from his rockets and artillery. Prince Abdul Calie and Jervas, meanwhile, understand that the technology is intended for times of war *and* peace—that, ultimately, it is a tool for enlightened communication that transcends cultural barriers, as Edgeworth and her father had earlier claimed. In inscribing the telegraph with Jervas's values and suggesting that Tippoo misuses the invention, Edgeworth renders this technology safe, legible, and moral. The novella thus assuages fears that the telegraph would enable a physical and ideological French invasion of Britain—fears that persist in the present as with the controversy about Huawei—by ultimately suggesting that literature has the power to determine the meanings, and possibly even the uses, of technologies.

NOTES

1. Matthew S. Schwartz, "Huawei Sues U.S. after Congress Bans Government Purchase of Its Equipment," NPR, March 7, 2019, https://www.npr.org/2019/03/07/700989603/huawei-sues-u-s-after-congress-bans-government-purchase-of-its-equipment.

2. John S. McCain National Defense Authorization Act for Fiscal Year 2019, Public Law 115-232, 115th Congress (Washington, DC: U.S. Government Publishing Office, August 13, 2018).

3. Ian Black, "NSA Spying Scandal: What We Have Learned," *The Guardian*, June 10, 2013, https://www.theguardian.com/world/2013/jun/10/nsa-spying-scandal-what-we-have-learned.

4. Stephanie Kirchgaessner, Paul Lewis, David Pegg, Sam Cutler, Nina Lakhani, and Michael Safi, "Revealed: Leak Uncovers Global Abuse of Cyber-Surveillance Weapon," *The Guardian*, July 18, 2021, https://www.theguardian.com/world/2021/jul/18/revealed-leak-uncovers-global-abuse-of-cyber-surveillance-weapon-nso-group-pegasus.

5. See Mary Fairclough, "The *Telegraph*: Radical Transmission in the 1790s," *Eighteenth-Century Life* 37, no. 2 (2013): 26–52; and Joanna Wharton, "Maria Edgeworth and the Telegraph," *European Romantic Review* 31, no. 6 (2020): 747–765.

6. As Aileen Douglas points out, "Lame Jervas" is loosely based on the life of William Smith, a student of the inventor and educator Andrew Bell. Douglas, *Work in Hand: Script, Print, and Writing, 1690–1840* (Oxford: Oxford University Press, 2017).

7. To distinguish between Maria Edgeworth and Richard Lovell Edgeworth, I refer to the former as "Edgeworth" and the latter as "Lovell Edgeworth." For a thorough account of Maria Edgeworth's role in co-authoring these essays with her father and the influence of the telegraph on her literary techniques, including those in "Lame Jervas," see Wharton, "Maria Edgeworth."

8. Geoffrey Wilson, *The Old Telegraphs* (London: Philimore, 1976), 12.

9. Russell Burns, *Communications: An International History of the Formative Years* (London: Institute of Electrical Engineers, 2004), 40.

10. Quoted in ibid., 42.

11. Ibid., 43.

12. *St. James's Chronicle or the British Evening Post* (London), September 18, 1794.

13. *Courier and Evening Gazette* (London), November 21, 1794; *Morning Chronicle*, November 25, 1794.

14. *Courier and Evening Gazette* (London), November 26, 1794.

15. *Morning Post and Fashionable World* (London), September 19, 1794.

16. Fairclough, "The *Telegraph*," 30.

17. Charles Dibdin, *A Collection of Songs, Selected from the Works of Mr. Dibdin. To Which Are Added, the Newest and Most Favourite American Patriotic Songs* (Philadelphia: Printed by J. Bioren for H. & P. Price, and sold by J. Rice, Baltimore, 1799), 255.

18. *Gentleman's Magazine* 64 (November 1794): 1175–1177. For a thorough discussion of the telegraph in the *Gentleman's Magazine*, see Fairclough, "The *Telegraph*," 30–31.

19. *Whitehall Evening Post* (London), September 27, 1794.

20. *Morning Post* (London), November 29, 1794.

21. Wharton, "Maria Edgeworth," 750.

22. Richard Lovell Edgeworth, *A Letter to the Right Hon. The Earl of Charlemont, on the Tellograph, and on the Defence of Ireland* (London: Printed for J. Johnson, in St. Paul's Church Yard, 1797).

23. Richard Lovell Edgeworth, "An Essay on the Art of Conveying Secret and Swift Intelligence," *Transactions of the Royal Irish Academy* 6 (1797): 95–139.

24. Lovell Edgeworth, *A Letter to the Right Hon. The Earl of Charlemont*, 4.

25. Maria Edgeworth and Richard Lovell Edgeworth, *Memoirs of Richard Lovell Edgeworth, Esq.* (London: Printed for R. Hunter, 1820), 358.

26. Lovell Edgeworth, "Secret and Swift Intelligence," 95.

27. Ibid., 104–105.

28. Ibid., 125.

29. Ibid., 100.

30. Ibid., 124.

31. Ibid., 116.

32. Maria Edgeworth, *Popular Tales* (London: Printed for J. Johnson, St. Paul's Churchyard, by C. Mercier and Co., 1804), 30.

33. Ibid., 31.

34. Ibid.

35. Ibid., 36.

36. Ibid., 48.

37. Ibid., 63.

38. Ibid., 72.

39. Ibid.

40. Kate Teltscher, *India Inscribed: European and British Writing on India, 1600–1800* (Delhi: Oxford University Press, 1995), 119.

41. Ibid.

42. Ibid., 111.

43. Ibid., 249.

44. Edgeworth, *Popular Tales*, 75.
45. Ibid., 75–76.
46. Ibid., 88–89.
47. Teltscher, *India Inscribed*, 249.
48. Wharton, "Maria Edgeworth," 758.
49. Ashley Cohen, "Wage Slavery, Oriental Despotism, and Global Labor: Management in Maria Edgeworth's *Popular Tales*," *The Eighteenth Century* 55, nos. 2–3 (2014): 193.
50. Edgeworth, *Popular Tales*, 89–90.
51. Sharon Alker, "Explosive Potential: Radicals and Chemistry in Maria Edgeworth's Fiction," *European Romantic Review* 22, no. 1 (February 2011): 14. Similarly, Wharton argues that "while invested in an Enlightenment ideology of 'improvement,' Edgeworth betrays a degree of caution about the democratization and globalization of scientific knowledge at a time of crisis in Europe" ("Maria Edgeworth," 749).

CHAPTER 8

Percy Shelley, Political Machines, and the Prehistory of the Postliberal

Jamison Kantor

I

One of the most celebrated accounts of the 2008 global economic collapse came from the former finance journalist and nonfiction writer Michael Lewis. The book was called *The Big Short* (2010) and, as is typical with Lewis's writing, it featured protagonists who, by escaping the inertia of the systems in which they had previously operated, end up as roguish heroes.[1] Despite the fanfare surrounding *The Big Short*, most respondents overlooked its original subtitle: *Inside the Doomsday Machine*. Throughout, Lewis describes the tortuous matrix of collateralized debt obligations (CDOs), derivatives, and credit default swaps as an integrated mechanism whose function is unconstrained by any single one of its operators. At various points throughout the text, the subprime mortgage lending industry against which the protagonists place their bets is called a "fantastic money machine," "a machine that turned lead into gold," and "an unthinking machine that could not stop itself."[2] The conceit that Lewis uses to describe finance capital—the fantastical automaton that has broken away from its masters, that is unconstrained from intention (whether singular or collective), and that is pushing history inexorably toward a cataclysmic reckoning—is not limited to his work alone. In their account of the financial crisis, Bethany McLean and Joe Nocera called the development of CDOs "a self-perpetuating machine," while the epilogue of their book is labeled, fittingly, "Rage at the Machine."[3] Evoking the apocalyptic (as well as literary) register oftentimes used to describe contemporary high finance, McLean and Nocera's volume was titled *All the Devils Are Here* (2010), a nod to Ariel's description of the calamitous shipwreck in *The Tempest* (1611).[4]

Scholars of eighteenth- and nineteenth-century literature have long been interested in the history of this rhetoric. Having tracked the metaphors of finance—or "fictitious capital"—from their affiliation with mass mechanical reproduction, immateriality and ghostliness, and even gothic narratives (which are replete with the kind of fatalism that the stock market seems to thrust on its participants), the past fifteen years of literary criticism has been fruitful for understanding the cultural forebears of the current regime of transatlantic capitalist accumulation.[5] These descriptions align with the traits of such a regime: labor is detached from capital systematically so that only people with money can make money, investment takes on an ever more abstract (or immaterial) quality in an attempt to mitigate risk, and a rising share of the national economy goes to investment vehicles held by a disproportionately small group of investors, leaving the rest of the population to the whims of a speculative market in which they cannot participate. In this essay, however, I propose a different, but corresponding, hypothesis about the aesthetics of social conditions in Europe and North America today. While critics have focused on the idiom of the self-governing, paranormal machine—or automaton—to describe the emergent finance economy, here I pivot to the modern state, the political formation that has always been tied to finance and the development of real industrial machinery and whose cultural depictions seem to inspire similar representations.

In classic political theory, the state—especially as it incorporated the ideas of republicanism and popular representation—is regularly compared to a machine or mechanism.[6] Thomas Hobbes likens the Leviathan, for instance, to an automaton whose motion springs from the consent of the people that it represents, while, in *The Social Contract* (1762), Jean-Jacques Rousseau argues that "the political machine" of the state gets its energy in more galvanic ways, as the sovereign taps into the "general will" of the people so as to "legitimise civil undertakings."[7] By the nineteenth century, the metaphor was in common parlance and was oftentimes used as a way to disparage unresponsive, "inorganic" leadership. Following Napoleon's defeat at Waterloo, the *Morning Post* reported that the emperor's "most ardent admirers and supporters, . . . convinced that nothing less than a miracle can restore their idol again to their embraces, will affect to consider him as an automaton, by whose motions they formerly were attracted, but whose springs are now disjointed and destroyed."[8] But if the imagined order of the liberal state had been signified traditionally as a machine energized by the constituency that it serves, I want to pose a critical question related to our current, perilous moment: What happens when this political machine starts to exclude those whom it is supposed to represent, shedding the need for representation and transforming into a self-contained, self-perpetuating apparatus? Like the "spectral" value produced by finance capital, will a modern state that has moved past the traditions of liberalism continue to exude the penumbra—or residue—of legitimacy without doing anything substantive to respond to the general will?

I advance this inquiry by turning to one of the most daring aesthetic, scientific, and political minds of the nineteenth century, Percy Bysshe Shelley. Known, among other things, for his influential protest lyrics—and for his pronouncement that "poets are the unacknowledged legislators of the World"—Shelley and his writing have been defined through a similar terminology of mechanism and spectral energy.[9] Of Shelley's final lyric fragment, *The Triumph of Life* (1822), William Hazlitt wrote that the poem was "full of fancy and of fire, with glowing allusions and wild machinery."[10] A comparable rhetoric of machinery and *techne* inhabits perhaps the most influential contemporary essay on the poet, Paul De Man's "Shelley Disfigured." For De Man, the "mechanical operation" of Shelley's lyricism reveals the limits of intentionality within any text, limits against which Shelley seemed to react.[11] As Marc Redfield and Orrin Wang have shown, the arbitrary nature of signification in Shelley—of "the mechanical iteration of inscription"—suggests the futility of social change but also the chance for a radical revision of the future.[12] This dialectic in Shelley's poetry between mechanism and the antimechanical thus articulates a kind of politics. Indeed, the range of possible signification within the poetry ratifies the broad spectrum of political ideology wherein Shelley has been located: for some critics, the poet is an enlightened aristocrat; for others, he is a cautious reformist; and still others consider him a radical proto-Marxist.[13]

Given the huge range of Shelley's insights on public life, I argue that a pair of later poems—*The Mask of Anarchy* (1819) and *The Triumph of Life*—registers the phenomenon of the political machine in three ways. First, the poems highlight Shelley's fascination with actual mechanical engineering—especially the construction of steam-powered vehicles—as a science that could harness the power of perpetual motion, or movement without human involvement. Second, the poems amalgamate Shelley's attraction to modern vehicular technology with his critique of the idea of "state machinery" and its tenuous connection to the people who give it legitimacy. Here, I turn to a specific genre: the masque. Often linked to the early modern period, the masque represented civic virtues in a kind of mechanical pageant. Featuring an automatic procession of allegorical figures signifying principles such as love, justice, and law, the masque authorized the abstraction of the state by exhibiting its personalized, humanistic nature to the public, which might be considered "happy passengers" on the moving vehicle of sovereignty. However, if the poem can be read as a diabolic parody—or an antimasque, as a host of scholars have suggested—then it also heralds the rise of the dominant regime of modernity, liberal republicanism. Glorified a decade later by supporters of the Reform Bill, *The Mask of Anarchy* envisions what Jeremy Bentham in 1817 called "the machine of government," a colossal, advancing apparatus that simultaneously represents and disciplines its subjects, incorporating them into its own instrumental logic.[14] However, in my third move, I argue that *The Triumph of Life*, the more speculative of both lyrics, proposes a stage of governance that may be more advanced than any under that we have lived, what I dub the "postliberal." Shelley's final poem foresees the separation of this governing vehicle—whether it represents or disciplines its subjects—from the people over whom it is supposed to preside. Nevertheless, the postliberal state continues to project a media spectacle of governance whose legitimacy becomes unimportant to the governed. At the risk of undue complexity, I argue that these lyrics stage simultaneously these many temporalities: the archaic world of the courtly masque, a near future governed by liberal institutions, and the distant future when the liberal state has been abandoned. Indeed,

Shelley's flattening of historical signification into formal shapes—what Tilottama Rajan has called "the reconstruction of appearances"—is a well-recognized hallmark of his poetics.[15]

This reduction of temporal categories into spatial ones brings to mind Fredric Jameson's renowned thesis in *Postmodernism* (1991).[16] But Jameson's more recent remarks about late capitalism are most applicable here: instead of concentrating on the exploitation of workers in modern Western economies, Jameson says that we should attend to their exclusion from those economies, as the unemployed, or "industrial reserve army of labor," come to constitute more grist for capital's churning mill.[17] In conjunction with our daily economic reality, this essay posits a new political concept: the state of exclusion, wherein formerly representational systems shed the capacity to respond to citizens in a substantive way. (My formulation is the obverse of Carl Schmitt's "state of exception," where the sovereign retains the privilege to act outside the rule of law). However, as is the case with all Shelleyan dichotomies, if there is darkness, there is also light. At the end, I lay out three potential paths forward detailed by *The Triumph of Life*: a reversion to civics based on pure aestheticism, an interregnum when people abandon a feckless political system and concentrate merely on their own livelihoods, and a utopian future when a refashioned political machine fades into the background and allows us to pursue higher goals for our lives.

II

For most contemporary readers, the epithet "political machine" recalls the end of the nineteenth century, not the beginning. The phrase evokes an ornate system based on patronage and hierarchy, a convoluted bureaucracy that can be circumvented through insider knowledge and, above all, the free flow of money into campaign coffers. The political machines of the United States, for instance, were infamous for influencing electoral politics and partisan affiliations. In New York City, the Tammany Hall organization—and its notorious leader, Boss Tweed—played an outsized role in developing the union and antimonopolist vote during the Gilded Age.[18] On the other side, the Republican "whiskey ring" centered in St. Louis during the 1870s employed machine politics for the purposes of tax evasion.[19]

However, in Europe at the turn of the nineteenth century, the concept of the political machine was broader and, perhaps, more fluid, as it moved across ideologies and featured in the writings of both radicals and reactionaries. At the conclusion of *The Rights of Man* (1791), Thomas Paine invoked the Enlightenment notion of a mechanism infused with metaphysical energy, declaring that the vast "machinery of Government" could be moved by two competing spirits: Reason or Ignorance.[20] Just two years earlier, however, Jean-Paul Marat eschewed this theoretical version of the machine. In an instance of revolutionary realpolitik, Marat proclaimed, "the political machine can only be wound up by violence, just as the air can only be cleared by a storm."[21] If Marat prefigures the guillotine—and Napoleon's march through Europe—in the first part of his analogy, then the second part revises the Enlightenment faith in reason as a guide to civil affairs. It is not the spirit of reason that fuels the machine of political modernity but instead the unstoppable motion of nature, which Marat—consummate journalist of the revolution—would subsequently liken to mass media, a similarly immaterial force that can "clear the air" and guide public opinion. "Do not be alarmed by words," pronounced Marat; "it is only by force that we can make liberty triumph and assure Public Safety."[22] As pamphlets and reports flew back and forth across Europe in the 1790s, a savvy rhetorician like Edmund Burke could thus be praised for knowing "more of the political machine than a recluse philosopher," as Hazlitt put it, and for "speculat[ing] more profoundly on its principles and general results than a mere politician."[23] Less interested in the principles and customs of the state than in its empirical design, Bentham conceived of government in 1817 as an apparatus that, like the animal body, must constantly remain in motion, for "when motion ceases, the body dies."[24] To keep the biomechanical state from "falling to pieces," there must be ceaseless, dynamic interaction between its representatives and their interests.[25] One generation after Burke, Bentham theorized that a legislature could be designed as a self-regulating colossus. But it must avoid behaving as "an automaton for signing papers," lest it invite corruption by appropriating for itself increasingly more money from the public trust.[26]

However, this interconnected apparatus of civic institutions, public opinion, and practical activity struck some commenters as artificial and restrictive. Offering a classic Romantic critique of mechanism, a young G.W.F. Hegel had called for citizens—"free human beings [treated] as if they were cogs in a machine"—to transcend the state entirely, a directive appropriated fifty years later by Karl Marx.[27] But Hegel's radicalism did not last. After Waterloo, the philosopher anticipated the "realist" liberalism that would eventually emerge as the dominant social structure, arguing in *The Philosophy of Right* (1820) that the political machine should merely be revised into a "system of the ethical world" that concretized the "rational life" of citizens as a collective.[28]

Of course, if Hegel urged citizens to accept the state and infuse it with new vitality, then he merely replaced a mechanical version of the political apparatus with an organic one.[29] Marx, in just one example of his manifold inversions of Hegel, noted in the final pages of *The Eighteenth Brumaire* (1852) that "all revolutions perfected this machine instead of breaking it."[30] While this was a tragic pronouncement on bourgeois society after the failure of the 1848 revolutions, it was also perhaps an indictment of Hegel's promotion of a single teleology of history. For Marx, the political machine was not just a structural formation but also an epochal one, a vehicle for one unvarying timeline of human development. Indeed, by the mid-nineteenth century, the concept of the perpetually moving state was not just affiliated with a Benthamite dance of interests.[31] It was also taking on a real, material appearance in the shape of mass transportation, especially the railways, a monumental technical achievement that was, in part, made possible by the state.[32] For the philosopher, the once-abstract state machine could be literalized by growing mechanical (or mechanized) national infrastructure: the state, in its pretenses to represent the masses, "snatched [common interests] from . . . society's members . . . and made [them objects] of government activity, . . . the railways, the national wealth and the national University of France."[33] In other words, Marx characterizes nineteenth-century liberal republicanism as a self-directed, autonomous vehicle that traverses one unvarying path of human history, remaining constant even when it is confronted by the passengers it carries. Both a concept and a concrete phenomenon, Marx's perpetual machine of state combines the technological advancement of publicly financed moving vehicles with a long-standing metaphor for the liberal political order. Indeed, it is both the civic metaphors and material reality of Europe that Marx wants to transform when, in the introduction of the essay, he calls for a "poetry . . . from the future" that will supersede the revolutionary language of days gone by.[34]

Indeed, the poet, if not the poetry, of the future may have been in the past. Writing more than thirty years before *The Eighteenth Brumaire*, Shelley had taken a keen interest in the industrial science of transportation and—in the syncretism typical of the polymath poet—how machines were inevitably influenced by the political milieu in which they were designed. No stranger to autonomous vehicles either, for years Shelley had represented in his poetry the self-propelled voyager as an emblem for politics and political awareness. In 1813's *Queen Mab*, for example, a flaming, flying chariot pilots the speaker across the globe, allowing him an aerial view of global oppression. In *Prometheus Unbound*, the closet drama that Shelley completed right before *The Mask*, the winged, "moonlike" "car of the hour," which conflates time with space, shuttles mythical figures to a decisive, revolutionary conflict.[35] During this period, Shelley was also exchanging letters with Henry Reveley—an engineer and friend of the family—about a prototype steamboat. Described by the poet in November 1819 as a "prosperous & swift voyager," the boat had promisingly incorporated a "steam cylinder and air-pump," components that Shelley commissioned from Reveley.[36] So inspired was the poet by transportation technology that, while writing in Reveley's lab one day, he linked poetic consciousness itself to an automated engine, "the self-impelling steam-wheels of the mind."[37]

Although Shelley's boat never entered the water (and the legacy of fatal maritime accidents continues to haunt any account of his life), a similar obsession with vehicular mechanism—and its ties to a "state machinery" that synthesizes institutions, public representation, and a specific track of progress—pervades *The Mask of Anarchy*. As one of the most influential protest poems in the history of British literature, *The Mask* was composed in the wake of the Peterloo Massacre of 1819. During this notorious event, regular and yeoman cavalry rode down labor protestors at Manchester's St. Peter's Square, killing six and wounding eighty. Shelley's poem interprets the event as a procession of four diabolical authorities—Murder, Fraud, Hypocrisy, and Anarchy—who ride into London, subjugate citizens and institutions to their will, and trample "the prostrate multitude" that dares to oppose them.[38] They are only stopped when an unnamed and nebulous "Shape arrayed in mail" emerges, inspiring protesters to put their bodies on the line in an act of nonviolent resistance, so that officials, potentially, "return with shame / To the place from which they came."[39] Composed with basic couplets and tight, four- to five-line stanzas of irregular tetrameter verse, the poem was a self-consciously popular text, one intended to compete with the numerous media representations of the event that had influenced its composition. (Due to legal concerns, however, its publication was withheld until 1832, when the poem was subsequently championed by supporters of the Reform Bill.)[40] In addition, the repetition of key refrains uttered by the authorities ("I AM GOD, AND KING, AND LAW") and the Shape of dissent ("Rise like lions after slumber. . . . Ye are many—they are few") function as mnemonic devices, apothegms that automate thought for readers.[41] Indeed, the professional classes depicted in the beginning of the poem echo the authorities' message mindlessly, "Whispering—'thou art Law and God.'"[42]

While the formal details of *The Mask* have a kind of mechanical regularity, its genre also conveys mechanism and systematic motion. Originally, the masque represented a civil pageant, one intended to personify the governing ideas of society as they descended from the sovereign and to present those ideas to the public in a routinized, but nevertheless engaging, order.[43] Typically associated with courtly life in the early modern period, the masque had a negative counterpart too, the aptly named "antimasque." Here, audiences were presented with the embodiment of rebellious or antisocial qualities that were subsequently regulated back into social propriety by the sovereign.[44] During the rise of the modern state, both masque and antimasque connected spectacle to moral instruction in order to materialize legitimacy through a distinctly populist aesthetic: the hierarchical control of the sovereign is made horizontal, diffused across different actors who work in concert with one another to support his/her claim to authority. Shelley's *Mask* reproduces the distinctly "lateral" quality of both genres by following its riders along a panorama of destruction, until the Shape finally halts them and instructs the people to disobey their rule. If the poem inverts the traditional masque, it also reproduces, as Amanda Jo Goldstein has noted, a genre that was even more current to the period: eighteenth-century didactic poetry in the neo-Lucretian tradition.[45] In these texts, the production of "natural histories" accounts for the deep, materialist dimension of political events that are unacknowledged by mere textual or iconographic recapitulation. However, if natural history "amplif[ies] the sensory and affective impacts" that the masque delivers through artifice and pageantry, then both genres rely on a similar representational schema: allegory, or the analogizing of one thing to another through a "degree of automaticism," to use De Man's famous phrase.[46] This imposition of allegory onto historical events turns what was incidental and contingent into a systematic operation of meaning, an engine that regiments the pieces and narrative of public affairs. Even the ethereal Shape—who is effervescent "as the light of sunny rain" and describes the populace as spectral "heroes of unwritten story"—could ultimately signify something purposive: the Hegelian "cunning of reason" working its way through the body politic in a decisive teleology of liberation.[47]

The iconography of the Shape is also distinctly contemporary. It appears at first "as a mist" and a "vapor of a vale," is born aloft by the gaseous energy with which

Shelley was fascinated, and passes over the heads of the multitude "as wind . . . so fast That they knew the presence there, / And looked,—but all was empty air."[48] Not just a distant object or divine totem, the Shape also urges onlookers to share its form, specifically by instructing them to view the decomposing bodies of the massacre as a kind of civic fossil fuel that will eventually power their movement: soon, says the Shape, "that slaughter to the Nation, / Shall steam up like inspiration."[49] As a supernatural technology of social change, the Shape might even resemble an early locomotive: it refigures the heat, steam, and, metal of Shelley's other imaginary vehicles—whether in his poetry or in contemporary prototypes—into an icon of populist resistance. And, in its blurry, dynamic motion above the heads of the people, the Shape suggests a perpetual future politics, in which, if we follow its example, history might advance endlessly. Drawing on masque, allegory, and contemporary engineering, the poem aestheticizes the idea of mechanism employed by postrevolutionary political theory. It is thus a fitting coincidence that England's first railroad—which opened in 1829, a decade after the publication of *The Mask*—connected Manchester with its fellow industrial hub, Liverpool.[50] As Paul Fyfe argues, by giving passengers a view of the "disparity in political representation," this route may have inspired support for the Reform Bill, which passed in Parliament twenty-one months later.[51]

To put this another way, *The Mask*, despite its futurist imagery, suggests a mode of governance that was evolving in Shelley's day: liberal republicanism, in which citizens are incorporated into a structure of representation that, presumably, responds to their interests. Critics have addressed Shelley's interpretation of Peterloo as, in Andrew Stauffer's words, "a contrast for harmonious visions of reconciliation" between the rulers and the ruled.[52] This is a fitting description for a protest song in which those rulers devastate their own people (and in which those people can only use this devastation—the strategy of nonviolent resistance—as a tactic for reform). But the line between consent and disorder is blurry here. The figures that ride in to quell dissent are not immediately known by their true names but are themselves clothed in the garb of authority: Murder "had a mask like Castlereagh" (the Tory minister of Parliament who had suppressed rebellion abroad), Fraud was "like Eldon" (the Lord Chancellor responsible for separating Shelley from his children), Hypocrisy appears "like Sidmouth" (the secretary of internal affairs known for domestic pacification), and Anarchy wears "a kingly crown / And in his grasp a scepter shone."[53] As they ride "through England proud and gay," they are greeted by an "adoring multitude."[54] And when the four figures finally descend on London, the various representatives of state administration—"a motley crowd" of soldiers, priests, and lawyers—bow "to the earth their pale brows" and afford them access to major institutions, including the Bank of England, Parliament, and the Tower of London, that venerable symbol of incarceration.[55] The representatives are so arrayed in "glorious triumph" that their true nature is concealed "even to the eyes," a phrase that conveys the inability of most people to see the duplicity of their leaders and, perhaps, the leaders' self-delusion. Indeed, it has become a truism to say that one of the many meanings of "mask" in Shelley's poem is the thin mask of liberal modernity worn by the agents of its corruption.[56]

These disguised figures evoke a well-known critique of bourgeois liberalism from the twentieth century that was also premised on the idea of political machinery: Louis Althusser's industrial and repressive state apparatuses, which regulate the practices of citizens through the "soft suppression" of voluntary institutions.[57] In other words, *The Mask* manifests a style of ideology critique in which consent and domination are seen as indistinguishable from each other; it is exemplary of what Jerrold Hogle describes as Shelley's ability to signify the transference of "past actions or ideologies to predetermine or restrain present modifications while suppressing the actual basis of outdated systems."[58] And yet the common people—embodied by the speaking Shape—are already aware of the repressive nature of the administrative structures that supposedly benefit them. The solution recommended by the Shape in a speech that constitutes the second half of the poem is an appeal to "the old laws of England" that were originally established by "children of a wiser day," to see themselves as "heirs of Glory," or the entailed inheritors of England's original Glorious Revolution, and to insist on nonviolent resistance, a tactic that can activate a sense of shame in those who read about the event and, hence, rebuild the circuit of affective nationalism.[59]

In short, the people must reform—not reject—the time-tested devices of representation. This can happen if they rescue from corruption an "original" liberalism, set it on the right trajectory once again, and sacrifice their bodies to revive a nationalist sentimentalism that was essential to Whig politics. Of course, the people's ability to refine the state and reincorporate their very bodies into it anticipates yet another twentieth-century criticism of liberalism, this time from Michel Foucault, who locates modern power, among other areas, in the rise of codified rationalism (governmentality) and the total management of corporeality (biopolitics).[60] And yet, even as the poem makes its readers aware of this new regime of modernity—in which phenomena of mass population ("Ye are many, they are few") and contiguous bodies ("Hand to hand, and foot to foot") require administration—it also prepares them to maintain and perfect such a regime.[61] This ability to improve the representational capacities of the liberal order has an aesthetic correlative: the population must don new, more accurate masks representing the old virtues of "Love, . . . Science, Poetry, . . . Thought, . . . Spirit, Patience, [and] Gentleness."[62] Hence, the masque—as a generic embodiment of the political machine—re-forms itself into a fresh pageant where abstract values are more tightly bound to their material signifiers. Even the most famous line in the poem (arguably the most famous line of Shelley's career), "Rise like lions after slumber," urges a garish metonymy between the common man and the Barbary lion, the most prolific symbol of English heraldry and, hence, national clout.[63]

By recovering conservative values and fitting them with new masks, the poem shows the possibility of the success of the representational state, not its failure. In doing so, Shelley dramatizes Bentham's 1817 plan to radically reform not disassemble, the state machine, just as he represents the "essential movement"—the single direction—of modern liberation described by Hegel one year later.[64] Alternatively, one could say that *The Mask* stages Marx's claim that all previous revolutions have simply refined the machine of politics. Indeed, when Shelley's publisher, Leigh Hunt, remarked that the poem would confront "the brazen masks of power" worn by the Tories, Shelley might have responded that the point of this confrontation was not to remove the masks but to make them less brazen and more authentic.[65]

The reformist heritage of the poem is alive and well. In 2017, the Labour Party manifesto featured the slogan "for the many, not the few," riffing on the lyric's central refrain, and proposed "rail electrification and expansion across the whole country," a callback to the poem's evocation of the emergence of the liberal state by, in part, its involvement with the machines of public transportation.[66] Astonishingly, Shelley's next masque pre-empts the debate over the efficacy of reformism and pushes it even further. *The Triumph of Life* intuits a radical future when the pretense of reformism is maintained without any substantive content. Pessimistic, intricate, and phantasmagoric, the lyric hypothesizes a situation in which the machines of modern politics whir faster and faster to stay in one place.

III

In 1821, one year before Shelley composed his *Triumph*, artists were still reacting to Peterloo and the emergence of a reactionary domestic agenda. Much like Shelley, they would invoke the image of an unstoppable mechanism of political oppression to advance their arguments. In a "partial revival" of Daniel Defoe's verse satire *De Jure Divino* (1706), the dissenter and satirist William Hone produced *The Right Divine of Kings to Govern Wrong* (1821), a sardonic poem that described all monarchs as indifferent automatons.[67] "The people like not his mechanic race," Hone wrote a year after the ascension of George IV to the throne. "They see no greatness in his youthful face."[68] (One suspects a dig here about the king's legendary fondness for luxury and fashion.)[69] Dedicated to the Holy Alliance—a coalition formed by Austria, Prussia, and Russia to stem the tide of secularism and liberal governance following the defeat of Napoleon in 1815—Hone's poem contained illustrations from the prolific cartoonist George Cruikshank. The title page, for instance, features a Saxon-looking king who stares into the foreground blankly, while priests anoint him with the "Oil of Steel" and the incense of "Discord," details that link the accouterments of divine right to the by-products of contemporary industrial engines: fossil fuel and steam. Adorned in the first image with a German *zweihander* sword and two nooses (figure 8.1), Cruikshank's king is even more mechanized and belligerent in a final cartoon, which depicts a lumbering iron giant (composed entirely of cannons and shot) who wields a torch and a sword to cut down radicals, signified by a Phrygian cap mounted on a pike (figure 8.2). In this image, however, the *zweihander* has been replaced with a British cavalry saber, and the Saxon war helmet is switched out for an English crown. This

THE

RIGHT DIVINE OF KINGS TO
GOVERN WRONG!

Dedicated to the Holy Alliance

BY THE AUTHOR OF
THE POLITICAL HOUSE THAT JACK BUILT.

" The devil will not have me damn'd, lest the *oil* that is in me should set hell on fire."
SHAKSPEARE.

LONDON:
PRINTED FOR WILLIAM HONE,
45, LUDGATE-HILL.

1821.

Eighteenpence.

Figure 8.1. The title page of William Hone's *The Right Divine of Kings to Govern Wrong!*, featuring a cartoon from George Cruikshank.

60 THE RIGHT DIVINE OF KINGS, &c.

No! He has issued no such foul command,
But dooms down Despots by the People's hand;
Marks tyrants out for fall in every age,
Directs the justice of the people's rage;
And hurling vengeance on all royal crimes,
Ordains the Revolutions of the times!

A thing of no bowels———
——— from the crown to the toe, topfull
Of direst cruelty.—His Realm a slaughter-house—
The swords of soldiers are his teeth——
Iron for Naples, hid with English *gilt*.

Shakspeare.

The End.

Printed by W. Hone, Ludgate Hill, London.

Figure 8.2. The final page of Hone's satire, with another cartoon from Cruikshank.

automaton sovereign no longer requires divine sanction for its power. It is now blessed by British industrial might.

Cruikshank's images express a convention regularly found in Shelley's later poetry and poetics: the old, even mythological, forms of tyranny reappear in the guise of modern industrial (or instrumentalist) rationality—a singular force that has nevertheless been separated from any one source or intention. Today, the critique of the instrumentality of culture is most often associated with the Frankfurt School, especially Theodor Adorno, whose legacy redounds in the work of contemporary Shelleyans. According to scholars, the poet prefigures Adorno's contribution to post-Marxist critique: Shelley is skeptical of the calculating instinct often associated with Enlightenment radicalism, but, like Adorno, he also posits a radical aesthetics that can move beyond instrumentality.[70] This sensitivity to the influence of mechanism is not just discursive but also appears at the level of the trope. Nearly fifty years before De Man noted the automated quality of Shelley's verse, William Empson perceived that the poet's work was characterized by "self-inwoven" similes, comparisons that refer ceaselessly to other places in the text—or to parts of the very simile in question.[71] Invoking the combination of divinity and machinery found in Hone's 1821 satire, Empson goes on to say that "*Form* [in Shelley's poetry] is its own justification; it sustains itself, like God, by the fact that it exists."[72] To put it another way, the lyric autonomy for which Shelley is known has regularly been connected to his trepidation with, or resignation to, automation and autoreferentiality.

Shelley's final poem, *The Triumph of Life*, proposes that such autoreferentiality—the process whereby things perpetuate themselves—could become the very model of modern politics in the future. Composed abroad in Italy after years of domestic upheaval in Britain, *The Triumph* is a highly abstract dream vision built on terza rima stanzas that are, in part, an homage to Dante's *Divine Comedy* (1320). It is also filled with imagery seemingly derived from Shelley's prior aesthetic engagements with liberal governance. As Shelley's autobiographical speaker slips into a "trance of wondrous thought," he encounters a blind, Janus-faced charioteer hurtling across the sky, a bleak version of Shelley's prolific motif of the automated vehicle as a symbol for unstoppable, unreflective progress.[73] Guided by the ghost of Rousseau, the speaker then encounters a series of historical spirits—Bacon, Voltaire, Kant, and Napoleon—chained to the car, a commentary on, among other things, the simultaneous inevitability and impotence of the Enlightenment project, which includes the domination of the natural world by modern science and technology.

These spirits are not alone, however. They are surrounded by a jubilant and "captive multitude" cheering on the procession of the charioteer as "some conqueror's advance."[74] Numbering a "million with fierce song and maniac dance," the multitude harks back to the supporters of tyrannical institutions featured in *The Mask of Anarchy*, now extrapolated to the greater populace.[75] Finally, after an encounter with the intangible figure of "shape all light"—perhaps a literalization of reason, as well as the dichotomy of creation and disruption that it presents—the lyric breaks off, ending in the middle of a response to an enigmatic question posed by the speaker: "Then, what is Life?"[76] Left unfinished due to Shelley's death, *The Triumph* is a classic Romantic fragment poem that keeps its readers suspended. Through its abrupt ending, it inspires the perpetual forward motion created by the piece's interlocking terza rima. Harking back to Cruikshank and Hone's *The Right Divine of Kings*, this description indicates a similar merger of industrial and divine activity found in representations of British tyranny during the Royalist revival following Waterloo.[77] But the chariot also drags a cheering, jubilant multitude along with it. Returning to the proto-locomotive imagery of *The Mask*—in which Shelley converted the steam power and perpetual motion of early transportation technology into a vehicular Shape that inspired the populace to revolt against injustice—the chariot now signals a diabolical inversion of political representation, where the people are bonded to their leaders by force, not by consent.

Even deep abstractions within the poem reconsider the semiotics of Shelley's previous poem. Revising the populist Shape from *The Mask*, who stops the "ghastly masquerade" in its tracks and provides the people with a template for future reform, is the "shape all light."[78] Unlike the solid Shape, the "shape all light" is ceaselessly protean, assuming a host of forms through the figure of simile—"like a willow," "as . . . slant morning beams that fell among the trees," "as . . . a shape of golden dew . . . Dances i' the wind where eagle never flew."[79] During its advance, the light hands the shade of Rousseau a "chrystal glass / Mantling with bright Nepenthe."[80] When Rousseau drinks, his memory is erased instantly. It should be noted that the "shape all light" is itself a trope of erasure. It not only blots out the phenomena that it touches but also deconstructs the figure of simile: when it moves, "all that was seemed as if it had been not."[81] Unlike *The Mask*'s Shape, which appears to the people as a break from the procession of tyranny—or perhaps as an event that, according to Alain Badiou, retroactively confers a new truth on a distinct historical occurrence—the "shape all light" is a perpetually mutable formation that moves, "silent as a ghost," alongside the revolutionary agent, leading him to yet another spectacle.[82] Rousseau sees the chariot again, but it is filled "with savage music," leading on "the loud million."[83] This time the multitude, "as if from some dread war / Triumphantly returning," drags along "a moving arch of victory"—the monument of revolutionary triumph now reconfigured as an engine of imperialist expansion—and then appears as atomies, floating together as parts of a single biomechanical war machine.[84] Through this sequence, the poem seems to stage the extreme dynamism of the idea of modern political progress—entailed by reform, revolution, or even reactionism—that has been tied, through images or rhetorical devices, to the concept of the automatic, or perpetually moving, apparatus. When we chain ourselves to these vehicles of representation, Shelley seems to argue, we must consent to the dialectical inversions that they might produce in their continuous, wheeling movement. In doing so, we are caught in an inescapable dialectic of modernity: the representatives of political progress always risk forgetting the conditions for which they once advocated and replicating the ones against which they originally stood.

However, another way to read the poem is as a vision of extreme stasis—a prophecy of a future when citizens merely *act* as if progress has continued unabated but when the representational mechanism has ceased to function, making any actual change futile. For instance, the lyric abounds with anaesthetized persons that participate in the "ghastly dance" of political history but who are, in effect, hollow, "shadows of shadows, yet unlike themselves."[85] Beyond the classic deconstructive message offered by Shelley's hollow figures, this emptying of signification conveys the futility of offering one's "second self"—one's *bios*, or civic identity—to a system of representation. In his seminal biography of the poet, Earl Wasserman notes that Shelley's poetry regularly featured such a "second self," conveyed oftentimes as a shadow or other phantom, who participates in the metaphysical world.[86] In *The Triumph*, however, that second self is extrapolated to an entire municipal crowd that has been denuded of independence: people walking on the "public way . . . [hasten] onward, yet none seemed to know / Whither he went, or whence he came, or why / He made one of the multitude."[87] Compelled to join the procession, the crowd is swept up in the chariot's track, but it has no say in its direction. Even the blindfolded driver cannot tell where the car will go.

Here, Shelley's extended conceit on the moving vehicle of state conveys its nonintentionality, a system of rule whose automatic processes exist beyond the consent of the governed but also beyond variation. Indeed, when Rousseau—who is one of the original members of this procession—exits the sky-pageant, he encounters the same panorama that the poet speaker sees in the first half of the poem. Even those who depart from the chariot's track are still compelled to look at the pageant hovering above them: when Rousseau is asked by the speaker in the penultimate lines of the poem to describe the essence of life after he has escaped, he "cast[s] / His eye upon the car which now had rolled / Onward, as if that look must be the last."[88]

The lyric's ability to dramatize the exclusion of subjects from a political machine of modernity that is both perpetual and unchanging intuits the postliberal. This governing order is realized in an even more distant future, when citizens are systematically shed from a conveyance of progress that no longer needs their representation to function but that also appears to endlessly, and futilely, recycle the same socioeconomic doxa. However, by concentrating on the constituents of this new contraption—and not on the lumbering, unstoppable Leviathan central to its original manifestations—*The Triumph* revises the aesthetics of the oppressive state machine featured in the Romantic era, as well as Carl Schmitt's pivotal twentieth-century idea of the state of exception, whereby the executive maintains power by removing himself from the institutions intrinsic to bourgeois liberalism.[89] In Shelley's lyric, these institutions are not superseded but, instead, have been transformed into sheer spectacle. Those who stand outside of institutions exist in a state of exclusion. They are stripped of representation but are nevertheless compelled to authorize this singular vision of modernity.[90]

Shelley's state of exclusion also seems to reconsider a major factor in previous representations of the political machine: the various forces—such as the divine right of kings, the cunning of reason, or the mass labor necessary for industrial development—binding the populace to the executive and driving them all forward. The lyric suggests that the compulsion to validate the new automaton of state—and to stare mutely at its continual turning—now comes from media spectacle, a phenomenon nearly literalized by the poem's procession of shadows, light, and grotesque pageantry. (Here, Shelley revisits Marat's key observation that, in modernity, public reason will be replaced by airy publicity.)[91] Among the many critics who have discussed the politics of spectacle in *The Triumph*, Joel Faflak comes closest to this analysis, stating that the lyric "predicts a history fascinated by the production of images that body forth life, but then compromise any return to life itself, a way to make the Empire live on, despite, or because of, its failure."[92] Connected to the compulsive expansion of capital into ever more ghostly shapes and abstractions, this government by spectacle, Faflak might say, responds to a population that has become endlessly addicted to the next hit of affect or scandalous pleasure—a fantasy that disseminates from an executive who is, in actuality, unable to steer history in any definitive direction.[93] However, if there are highs, then there are also comedowns. Citizens are made equally exhausted and cynical by this procession of signifiers and stare blankly, like the shade of Rousseau, at the masque(s) recurring before them. William Wordsworth described this phenomenon as the "savage torpor" produced by the emerging mass media of the 1790s, whose constant sensationalism blunted one's ability to rank or assess value.[94] A generation later, Shelley's lyric would amplify Wordsworth's remark and apply it to the condition of the *polis*: subjects find themselves suspended, unable to discern profound events from insignificant ones, and—much like the unanswered final question of the poem—waiting for an event that remains outside the prospect of the existing order of meaning.

In 1824, *The Triumph of Life* appeared for the first time as a selection in Shelley's *Posthumous Poems*. Describing Percy's most memorable qualities in the preface to the volume, Mary Shelley reports that her husband was "a bright vision, whose radiant track, left behind in the memory, is worth all the realities that society can afford."[95] Her reflections recall the predominant image of his final lyric: the charioteer streaking across a sky followed by a nebulous pageant. They also act as a tribute to the ubiquitous Platonism that animated Shelley's poetry. Beyond these qualities, however, Mary Shelley's observation registers the same deconstruction that characterizes her husband's lyrics. For it is at once a commemoration and a warning: while poetry offers us access to the "bright vision" that exists beyond the veil of the material world, it also mimics the fabulation for which that world might be easily traded.

IV

So what is life? More specifically, what will life resemble after the ascension of a post-liberal machine of politics? I want to reference the critical heritage of *The Triumph of Life*—which has elevated the poem as one of the major works of speculative literature in the Romantic period—and close by outlining three potential futures implied by the lyric: a return to a wholly aestheticized politics, a retreat from politics entirely and its replacement with insular groups that offer their own *Leitkultur*, and a utopian renewal of the concept of the political machine, one that entails a commitment to representation and affirms the emancipatory qualities that attend it.

The Triumph is a highly aesthetic performance. With its polychromatic texture, it invokes an amazing expanse of media history, from the sacred image to the written word to the cinema.[96] It is also, however, a meditation on the influences and inadequacies of aestheticism for living. Indeed, recent studies of the poem have tended to engage directly this dialectic between artistic absorption and skepticism: *The Triumph* is both a perfect example of the aesthetic ideology endemic to the greater Romantic lyric and—as it preempts critique from Hegel, Adorno, and De Man—a savvy commentary on the critical powers and limits of art; it is an excursus that happens to masquerade as art itself.[97] There is a third option, however, and it seems to capture the representation of subjects in the poem who attend compulsively—and wearily—to the pageant of civic life that passes before them, making "Circles around it like the clouds that swim / Round the high moon in a bright sea of air / And more did follow, with exulting hymn."[98] The lyric can be read as proposing a future when, given the complete detachment of subjects from governing institutions, the aesthetic may be the only possible choice left for mass engagement. (One must also note the simultaneous possibility here that, under late capitalism, this kind of art will be subsumed into the logic of the market completely. Shelley's poem also contains this augury: troubled by the continued proliferation of paper money and "the ghost of gold," the poet fills *The Triumph* with a series of translucent figurations that mimic the intangibility of abstract capital.)[99]

Not only would civics be replaced by frivolous media, but private citizens might become its greatest producers. By creating a dynamic simulacrum that stands in for degraded or nonexistent institutions, these endeavors would be part of what Adam Curtis has dubbed "hypernormalisation": the fabrication of a simplistic, functioning political order that obscures the chaotic reality lurking beneath.[100] In the current moment, it is right-wing internet provocateurs—or trolls—who seem to be most at home with the hypernormal, as they recycle political memes and screeds whose absurdist quality allows these texts to slip between irony and a kind of hyperbolic sincerity.[101] While the hyperreal texture, satiric quality, and glorious pageantry of *The Triumph* might appeal to postliberal internet reactionaries, its abstract qualities and the disinterested gaze of its poet-speaker gesture to the rise of a class of leftist bohemians—perhaps like the original Shelley-Byron cohort—who champion art because of its independence from hegemony, as well as its ability to expose, but not rectify, systemic absurdities. Commenting on the hipster enclaves emerging in urban centers, one economist (who never fails to naturalize capital) has said that a new, precarious bohemianism will be one of the outcomes of our acceptance of dangerous economic inequality.[102]

The second future entailed by the poem involves another retreat from the modern structure of liberal representation. But this timeline does not feature an attempt by citizens to reconstitute a lost public sphere through aesthetics. Instead, the lyric suggests an age marked by the endless deferral of any totalizing political order, an adjournment that Shelley's readers have seen in the conditional tone of his political verse and in the poet's commitment to the idea of messianic time.[103] One particular strength of *The Triumph* is its ability to illustrate this kind of temporal suspension: the blind charioteer of history does not move in any conclusive direction but instead follows arbitrarily the tracks of a hovering sun, a guide star that, paradoxically, announces the end of history, illuminating "all that is, has been, or will be done."[104] In the book *How Will Capitalism End?* (2016), the sociologist Wolfgang Streeck dubs this suspended future a "lasting *interregnum*." Streeck argues that, following the breakdown of the bourgeois liberal consensus, the next epoch would be "a prolonged period of social entropy."[105] Here, atomized groups and epistemologies would emerge to fill the vacuum left by a discredited order that nevertheless has fomented no serious counterorder. (In the lyric, the common people—hollowed into the form of shadows and suspended in the air—cannot join the train following the charioteer but "wheel . . . Round them and round each other, and fulfill / Their work," withdrawing into their everyday pursuits.)[106] The result is a replacement of "structure"—or the collective institutions that had previously made up various political machines—with "culture," where fragmented pockets of the population reformulate themselves through "fuzzy" custom, norms, or even ethnicity.[107]

The stuff that binds together the disparate human shapes in *The Triumph* is usually described as "life," a Platonic ideal that Shelley once named the "spirit within [man] at enmity with nothingness and dissolution."[108] While Shelleyan life has been cast as a vitalist force that resists oppressive social categories—heteronormativity, the hyperspectacle of late capital—one can also understand it, like *The Triumph*'s chariot, as a substance frozen in time, avoiding any confrontation in favor of its own limited activity.[109] In fact, this phenomenon may be visible in the practices of the contemporary academy, which, in large part due to its subjugation to a regime of metrics and utilitarian outcomes, has produced a kind of quietist materialism, where politics—or even a basic social urgency—are explicitly removed from empirical scholarship or scholarship about empiricism. This situation also recalls the poet's belief in, to reference Robert Mitchell's recent analysis of Shelleyan vitalism, suspended animation, where the "non-action" of the mind in the face of sociomaterialist upheaval inspires a variety of passive resistance, a cornerstone of liberal politics.[110] But how long must knowledge workers remain suspended? Oftentimes, the result has been an evolutionary theory without subjects, an Anthropocene without capitalism, and a new scientism that sometimes evades politics as a matter of duty.[111]

However, *The Triumph* does not just forecast the end of political praxis in modern life—or its withering on the vine of an effete liberalism. The lyric also hints at an even more distant future when the structure of politics has long since been reformed around the multitude. Furthermore, Shelley intimates that the operation of that structure will become so autonomous—so automatic—that the subjects it serves can, for the most part, forget that it exists.[112] At first, this utopian prediction seems unlikely. The lyric is bleak on precisely this issue; it is almost defined by its refusal to place images, tropes, and persons within a coherent representational schema, whether that schema is political, rhetorical, or both. And yet, within a central tableau of *The Triumph*, icons that had previously signified exclusion reverse themselves temporarily and offer a picture of unification. When the ghost of Rousseau has his own dream vision, he awakes to see a "common Sun" that "Sheds [light] on the common earth," a bid for the expansion of enlightenment beyond the boundaries of private control. Given Shelley's fascination with natural energy, perhaps the line is also a prophecy of communal solar power, spreading across the earth for all to use. The light from that sun burns "on the waters of a well that glowed / Like gold," an image that upends Shelley's pervasive fear of fictitious finance by converting matter that is merely "like gold" into a public resource. It illuminates "the orient cavern" in which Rousseau finds himself, a representation that revises Shelley's earlier allusion to Bacon and a scientific method guided by personal acquisition—where "lightning . . . wake and unbar the caves" of ignorance that hold "The treasure of the secrets of its reign"—into a kaleidoscopic "blaze" with many paths, a method of knowledge creation that has been unbound from a single political economic regime. And Rousseau's vision is not guided by Reason—which can be instrumentalized—but Love, a force that defies transfiguration.[113]

All these structural changes—the expanded commons, the democratization of raw materials and the productive apparatus, the development of knowledge projects liberated from socioeconomic hegemony, the society bonded together by care—have been proposed today.[114] All of them would require vigorous public effort to achieve. Ironically, we would know that this future has arrived, the lyric implies, when the structures on which it had been built take root within the depths of consciousness, in dreams that occur within dreams. To witness the triumph of this new political machine would be to see that triumph not as an arch of victory laid on the backs of the people but as a diaphanous beam of light that is both there and not there, as external guide and internal conviction—and under whose "obscure tenour" we are all covered.[115]

NOTES

1. As evidence of this claim, I note the 2015 film adaptation of *The Big Short*, which, at the time, featured no less than three of Hollywood's most notable leading men: Brad Pitt, Ryan Gosling, and Christian Bale. For the original book, see Michael Lewis, *The Big Short: Inside the Doomsday Machine* (New York: Norton, 2011).

2. Lewis, *The Big Short*, 70, 73, 151.

3. Bethany McClean and Joseph Nocera, *All the Devils Are Here: The Hidden History of the Financial Crisis* (New York: Portfolio/Penguin, 2011), 122, 358.

4. William Shakespeare, *The Tempest*, 1.2.335, OpenSource Shakespeare, accessed July 2019, https://www.opensourceshakespeare.org/views/plays/play_view.php?WorkID=tempest&Act=1&Scene=2&Scope=scene.

5. For the intersection between mass print and the paper money that underwrote high finance, see Mary Poovey, *Genres of the Credit Economy: Mediating Value in Eighteenth- and Nineteenth-Century Britain* (Chicago: University of Chicago Press, 2008), 153–169. For the spectral quality of finance, see Ian Baucom, *Specters of the Atlantic: Finance Capital, Slavery, and the Philosophy of History* (Durham, NC: Duke University Press, 2005), 141–169; Alex Dick, *Romanticism and the Gold Standard: Money, Literature, and Economic Debate in Britain, 1790–1830* (London: Palgrave Macmillan, 2013), 123–149. The connection of fate with financial speculation is discussed in Christian Thorne, "Providence in the Early Novel, or Accident If You Please," *MLQ* 64, no. 3 (2003): 323–347; and Jamison Kantor, "Horace Walpole and the Fate of Finance," *The Eighteenth Century* 58, no. 2 (2017): 135–155.

6. A long intellectual history of mechanism in political theory can be found in Giuseppa Saccaro-Battisti, "Changing Metaphors of Political Thought," *Journal of the History of Ideas* 44, no. 1 (1983): 31–54.

7. Thomas Hobbes, *Leviathan*, rev ed. (Toronto: Broadview Press, 2011), 1–2; Jean-Jacques Rousseau, *The Social Contract*, trans. G.D.H. Cole (New York: E. P. Dutton & Co., 1923), 18.

8. "Sailing of Bonaparte," *The Morning Post* (London), August 7, 1815.

9. Percy Bysshe Shelley, "A Defence of Poetry," in *Shelley's Poetry and Prose*, ed. Donald H. Reiman and Neil Fraistat (New York: Norton, 2002), 535.

10. William Hazlitt, "Review of Shelley's *Posthumous Poems*," in *Shelley: The Critical Heritage*, ed. James E. Barcus (London: Routledge and Kegan Paul, 1975), 341.

11. Paul De Man, "Shelley Disfigured," in *Deconstruction and Criticism*, ed. Harold Bloom (New York: Continuum, 2004), 37.

12. Marc Redfield, *The Politics of Aesthetics: Nationalism, Gender, Romanticism* (Stanford, CA: Stanford University Press, 2003), 159. See also Orrin N. C. Wang, *Fantastic Modernity: Dialectical Readings in Romanticism and Theory* (Baltimore: Johns Hopkins University Press, 1996). Wang reckons with Shelley's *The Triumph of Life* via De Man's "Shelley Disfigured," an essay that, according to Wang, illuminates the poem's portrayal of "a violently linguistic world in which the only action is the arbitrarily positing 'madness of words'" (*Fantastic Modernity*, 46).

13. For the view of Shelley as an "agrarian reactionary"—who wanted to preserve the vestiges of benevolent aristocracy—see Donald Reiman, "Shelley as Agrarian Reactionary," in Reiman and Fraistat, *Shelley's Poetry and Prose*, 589–599. Jerrold Hogle sees a poem such as *The Mask of Anarchy* as a bid to restore the true liberty associated with "'Constitutional' government," not a call to overcome that hegemonic system. Hogle, *Shelley's Process: Radical Transference and the Development of His Major Works* (New York: Oxford University Press, 1989), 138. Finally, for a definitive account of Shelley's pre-Marxist radicalism, see Paul Foot, *Red Shelley* (London: Sidgwick and Jackson, 1980).

14. Jeremy Bentham, *Plan of Parliamentary Reform* (London: R. Hunter, 1817), 52.

15. Tilottama Rajan, *Dark Interpreter: The Discourse of Romanticism* (Ithaca, NY: Cornell University Press, 1980), 70.

16. Fredric Jameson, *Postmodernism, or, The Cultural Logic of Late Capitalism* (Durham, NC: Duke University Press, 1991). As Jameson said in 1991, "the return to history everywhere remarked today demands closer scrutiny in the light of this 'historical' perspective—only it is not a return exactly, seeming rather to mean incorporating the 'raw material' of history and leaving its function out, a kind of flattening and appropriation" (325).

17. Jameson's *Representing Capital: A Commentary on Volume One* (London: Verso, 2011) is, in the author's own words, "a book about unemployment" (2). "Being unemployed or without economic function is no longer to be expelled from capital," as Jameson says, "but to remain within it. Where everything has been subsumed under capitalism, there is no longer anything outside it; and the unemployed . . . are as it were employed by capital to be unemployed; they fulfill an economic function by way of their very non-functioning" (71). Jameson's comment applies to low unemployment environments, too: contemporary capitalist economies will sometimes tame inflation by inducing recessions, thus driving up unemployment.

18. See Terry Golway, *Machine Made: Tammany Hall and the Creation of Modern American Politics* (New York: Norton, 2014), 134–141. For a fulsome treatment of the term "political

machine," see Adam T. Smith, *The Political Machine: Assembling Sovereignty in the Bronze Age Caucasus* (Princeton, NJ: Princeton University Press, 2015), 7–11, 27–58.

19. Robert C. Kennedy, "Probe Away," *HarpWeek*, accessed June 2019, https://www.harpweek.com/09Cartoon/BrowseByDateCartoon.asp?Month=March&Date=18.

20. Thomas Paine, *The Rights of Man* (London: J. S. Jordan, 1791), 161.

21. Marat quoted in Michel Vovelle, *The Fall of the French Monarchy, 1787–1792* (Cambridge: Cambridge University Press, 1984), 209.

22. Marat quoted in ibid. Contrast Marat's claims about the importance of mass media with theorists of the Atlantic Republican tradition, who find a relationship between rationality and artifice in the composition of political modernity. Michel Foucault, for instance, discusses the interdependence of the state—which "is governed according to rational principles that are intrinsic to it"—with the art of government, which "find[s] the principles of its rationality in . . . the specific reality of the state." Foucault, "Governmentality," in *Power: Essential Works of Foucault, 1954–84*, trans. Robert Hurley (New York: New Press, 2000), 213.

23. William Hazlitt, "Review of *Biographia Literaria*," *Edinburgh Review* 28 (1817): 505.

24. Bentham, *Plan of Parliamentary Reform*, 52.

25. Ibid., 229.

26. Ibid., 195.

27. G.W.F. Hegel, "The Oldest Systematic Programme of German Idealism," in *The Early Political Writings of the German Romantics*, ed. and trans. Frederick C. Baiser (Cambridge: Cambridge University Press, 1999), 4.

28. G.W.F. Hegel, *Hegel's Philosophy of Right*, trans. T. M. Knox (Oxford: Clarendon Press, 1978), 170.

29. See Jürgen Habermas, *The Philosophical Discourse of Modernity* (Cambridge, MA: MIT Press 1990), 27.

30. Karl Marx and Frederick Engels, *The Eighteenth Brumaire of Louis Bonaparte*, in *Collected Works*, vol. 11 (New York: International Publishers, 1979), 186.

31. Bentham, *Plan of Parliamentary Reform*, 229.

32. In his study on the cultural impact of the nineteenth-century railroad, Michael Freeman reports that the British Parliament "legislated the first public railways into existence," but, in the interest of efficiency and punctuality, allowed them to remain a de facto private monopoly. See Freeman, *Railways and the Victorian Imagination* (New Haven, CT: Yale University Press, 1999), 28.

33. Marx and Engels, *The Eighteenth Brumaire*, 186.

34. Ibid., 106.

35. Percy Bysshe Shelley, *Prometheus Unbound*, in Reiman and Fraistat, *Shelley's Poetry and Prose*, 267, 256.

36. Percy Bysshe Shelley to Henry Reveley, November 17, 1819, in *The Letters of Percy Bysshe Shelley*, vol. 2, *Shelley in Italy*, ed. Frederick L. Jones (Oxford: Oxford University Press, 1964), 158. For Ivan Ortiz, "poetic making is intimately linked to the figure of transport in Shelley's work," while "the steamboat (in particular) might be said to have represented not only a new kind of poetic figuration, which projects an inner ideal out into the world, but also the practical application of poetry for social and political reform." Ortiz, "Romanticism and the Aesthetics of Modern Transport" (PhD diss., Princeton University, 2013), 119, 158.

37. Percy Bysshe Shelley, "Letter to Maria Gisborne," in Reiman and Fraistat, *Shelley's Poetry and Prose*, 332.

38. Percy Bysshe Shelley, *The Mask of Anarchy*, in Reiman and Fraistat, *Shelley's Poetry and Prose*, 320.

39. Ibid., 319, 325.

40. Ibid., 315.

41. Ibid., 317, 320, 326.

42. Ibid., 318.

43. For a good overview of the civic pageantry of the early modern masque, see Michael Witmore, "Arrow, Acrobat and Phoenix: On Sense and Motion in English Civic Pageantry,"

in *The Automaton in English Renaissance Literature*, ed. Wendy Beth Hyman (Burlington, VT: Ashgate, 2011), 109–127.

44. Various critics have discussed the importance of the antimasque to Shelley's lyric art. See Stuart Curran, *Shelley's Annus Mirabilis: The Maturing of an Epic Vision* (San Marino, CA: Huntington Library Press, 1975), 190–191; Stephen Behrendt, *Shelley and His Audiences* (Lincoln: University of Nebraska Press, 1989), 201; Andrew Stauffer, *Anger, Revolution, and Romanticism* (Cambridge: Cambridge University Press, 2005), 15.

45. Amanda Jo Goldstein, *Sweet Science: Romantic Materialism and the New Logics of Life* (Chicago: University of Chicago Press, 2017), 167.

46. Ibid., 168; Paul De Man, *Blindness and Insight: Essays in the Rhetoric of Contemporary Criticism* (London: Routledge, 2005), 199.

47. Shelley, "*The Mask of Anarchy*," 319, 320. Goldstein sees the aestheticism of the Shape as a blatant imposition of rhetorical machinery. She describes its sudden appearance as a "*figura-ex-machina* moment in which a flagrantly poetic 'Shape' handily finishes off the task of political struggle" (*Sweet Science*, 167). Kir Kuiken, on the other hand, sees the Shape as the possibility of figuration itself and, hence, a sign of openness and hope for radical democratic reform. See Kuiken, *Imagined Sovereignties: Toward a New Political Romanticism* (New York: Fordham University Press, 2014), 196.

48. Shelley, *The Mask of Anarchy*, 319.

49. Ibid., 326.

50. For the social and aesthetic meaning of the Liverpool-Manchester Railway—on its opening day in September 1830—see Paul Fyfe, "On the Opening of the Liverpool and Manchester Railway, 1830," *BRANCH: Britain, Representation and Nineteenth-Century History*, December 2012, http://www.branchcollective.org/?ps_articles=paul-fyfe-on-the-opening-of-the-liverpool-and-manchester-railway-1830.

51. Ibid.

52. Stauffer, *Anger, Revolution, and Romanticism*, 15.

53. Shelley, *The Mask of Anarchy*, 316–317.

54. Ibid., 317.

55. Ibid., 318.

56. Ibid., 317.

57. Louis Althusser, "Ideology and Ideological State Apparatus (Notes towards an Investigation)," in *Lenin and Philosophy and Other Essays* (New York: Monthly Review Press, 2001), 85–126.

58. Hogle, *Shelley's Process*, 323.

59. Shelley, *The Mask of Anarchy*, 325, 320.

60. See Foucault, "Governmentality," 211–219; Michel Foucault, *The Birth of Biopolitics: Lectures at the Collège de France, 1978–79*, trans. Graham Burchell (London: Palgrave Macmillan, 2008), 51–73.

61. Shelley, *The Mask of Anarchy*, 327, 329.

62. Ibid., 323.

63. Ibid., 326.

64. Hegel, *Philosophy of Right*, 101.

65. Leigh Hunt, "Disturbances at Manchester," *The Examiner* (London), August 22, 1819, 530.

66. Labour Party, *For the Many, Not the Few: The Labour Party Manifesto 2017*, October 2017, 11, https://labour.org.uk/wp-content/uploads/2017/10/labour-manifesto-2017.pdf.

67. William Hone, *The Right Divine of Kings to Govern Wrong!* (London: William Hone, 1821). Hone and the cartoonist George Cruikshank had condemned the Peterloo Massacre two years earlier, collaborating on a brutal doggerel satire called *The Political House That Jack Built* (1819). See William Hone, *The Political House That Jack Built*, Romantic Circles: Editions, 1998, https://romantic-circles.org/editions/hone/jacktext.htm.

68. Hone, *The Right Divine of Kings*, 26.

69. Scholars have linked the prince regent—who ascended to the throne in 1821—to Frank Churchill, Jane Austen's self-obsessed dandy in *Emma* (which was, of course, dedicated to

the prince). See, for instance, Jocelyn Harris, *Satire, Celebrity, and Politics in Jane Austen* (Lewisburg, PA: Bucknell University Press, 2017), 189.

70. For Shelley's aversion to calculative rationality, see Robert Kaufman, "Legislators of the Post-Everything World: Shelley's 'Defence' of Adorno," *ELH* 63, no. 3 (1996): 714–725. For the limitless horizon of political change, see Matthew Borushko, "Aesthetics of Nonviolence: Shelley, Adorno, Rancière," in *The Politics of Shelley: History, Theory, Form*, Romantic Circles: Praxis Series, October 2015, https://www.rc.umd.edu/praxis/shelley_politics/praxis.2015.shelley_politics.borushko.html. This entire volume from *Romantic Circles* is worth consulting for links between Shelley and Adorno.

71. William Empson, *Seven Types of Ambiguity* (New York: New Directions, 1966), 160.

72. Ibid., 161.

73. Percy Bysshe Shelley, *The Triumph of Life*, in Reiman and Fraistat, *Shelley's Poetry and Prose*, 484.

74. Ibid., 487.

75. Ibid. London's population reached one million in 1800. Given Shelley's interest in Malthus's theory of population and nourishment, he may have been aware of this figure.

76. Ibid., 494, 500.

77. On the conservatism following 1815, see Marilyn Butler, *Romantics, Rebels, and Reactionaries: English Literature and Its Background, 1760–1830* (Oxford: Oxford University Press, 1985), 5–6; Kevin Gilmartin, *Writing against Revolution: Literary Conservatism in Britain, 1790–1832* (Cambridge: Cambridge University Press, 2006).

78. Shelley, *The Mask of Anarchy*, 317; Shelley, *The Triumph of Life*, 494.

79. Ibid., 494–495.

80. Ibid., 494.

81. Ibid., 495.

82. Ibid., 494, 497; Alain Badiou, *Being and Event*, trans. Oliver Feltham (New York: Continuum, 2007), 173–261.

83. Shelley, *The Triumph of Life*, 497.

84. Ibid. For a keen analysis of the ideological and architectural meanings of this monument, see Wang, *Fantastic Modernity*, 60–65.

85. Shelley, *The Triumph of Life*, 500, 498.

86. Earl Wasserman, *Shelley: A Critical Reading* (Baltimore: Johns Hopkins University Press, 1977), 9.

87. Shelley, *The Triumph of Life*, 484–485.

88. Ibid., 500.

89. Carl Schmitt, *Political Theology: Four Chapters on the Concept of Sovereignty*, trans. George Schwab (Chicago: University of Chicago Press, 2005), 5–35.

90. Wang, on the other hand, sees the spectacle of those institutions as disputing the permanence of any political legitimacy: "As a simulacrum, the discourse of the triumph becomes the perfect emblem of the French Revolution, as an event with no real truth or legitimacy behind the guises it steals from other centuries" (*Fantastic Modernity*, 60).

91. Marat quoted in Vovelle, *The Fall of the French Monarchy*, 209.

92. Joel Faflak, "Right to Romanticism," *European Romantic Review* 27, no. 3 (2016): 290.

93. Ibid., 287–289.

94. William Wordsworth, Preface to *Lyrical Ballads*, in *Lyrical Ballads, and Other Poems, 1797–1800* (Ithaca, NY: Cornell University Press, 1993), 746.

95. Mary Shelley, preface to *Posthumous Poems of Percy Bysshe Shelley* (London: John and Henry L. Hunt, 1824), iv.

96. For the cinematic ancestors of Shelley's simulacra in *The Triumph*, see Faflak, "Right to Romanticism," 289–293.

97. According to Rajan, *The Triumph* stages one of Hegel's major categories of art—the symbolic—by presenting a parade of ideas that resist stable material receptacles. See Tilottama Rajan, "The Work of the Negative: Symbolic, Gothic, and Romantic in Shelley and Hegel," *Studies in Romanticism* 52, no. 1 (2013): 21–28. For Shelley's indirect influence on

Adorno, aesthetic ideology, and the "'critical nature' of [the] modern aura," see Robert Kaufman, "The Work of Romanticism in the Age of Mechanical Postmodernism," *European Romantic Review* 12, no. 2 (2001): 239.

98. Shelley, *The Triumph of Life*, 497.

99. The "ghost of gold"—or the proliferating abstraction of capital—is Dick's formulation, and he takes the term from Shelley's *Mask* (321). See Dick, *Romanticism and the Gold Standard*, 110–149.

100. Adam Curtis, dir., *Hypernormalisation* (London: BBC, BBC iPlayer, 2016).

101. Redfield is very lucid on the dangers of Shelley's aestheticized politics, as well as the poet's deconstruction of the liberal order, an approach that could very well result in a future open to reactionaries, not radicals. See Redfield, *The Politics of Aesthetics*, 148–171.

102. The economist is Tyler Cowen. See his interview on *Morning Edition*: "Tired of Inequality? One Economist Says It'll Only Get Worse: Interview with Tyler Cowen," NPR, September 12, 2013, https://www.npr.org/transcripts/221425582.

103. James Chandler, for instance, argues for this form of historical time in *The Mask of Anarchy*, especially as the poem competed with print journalism to register the events that were unfolding: "[The poem] displaces the everyday time of the periodicals into a quasi-apocalyptic framework, anticipating the displacement of 'empty homogenous time' by 'messianic time' in Benjamin's twentieth-century analysis." Chandler, *England in 1819: The Politics of Literary Culture and the Case of Romantic Historicism* (Chicago: University of Chicago Press, 1998), 529. Concentrating on the conditional word "may," Susan Wolfson addresses Shelley's equivocations over politics in a classic reading of the sonnet "England in 1819." Wolfson, *Formal Charges: The Shaping of Poetry in British Romanticism* (Stanford, CA: Stanford University Press, 1997), 202–206.

104. Shelley, *The Triumph of Life*, 486.

105. Wolfgang Streeck, *How Will Capitalism End? Essays on a Failing System* (London: Verso, 2016), 13.

106. Shelley, *The Triumph of Life*, 488.

107. Streeck, *How Will Capitalism End?*, 37–46.

108. Percy Bysshe Shelley, "On Life," in Reiman and Fraistat, *Shelley's Poetry and Prose*, 506.

109. For Faflak, Shelleyan "life" registers the spectacle of capital whose payout is infinitely deferred—"a production of distinct images whose unfolding spectacle postpones distinction indefinitely"—and the traumatic kernel that resides within that fantasy. Joel Faflak, "Dancing in the Dark with Shelley," in *Constellations of a Contemporary Romanticism*, ed. Jacques Khalip and Forest Pyle (New York: Fordham University Press, 2016), 151. Lee Edelman takes Shelleyan "life" to have a "constitutive negativity" that offers an opening for subjects to reject reproduction—and the sexual categories influencing it—altogether. Edelman, "The Pathology of the Future, or the Endless Triumphs of Life," in Khalip and Pyle, *Constellations of a Contemporary Romanticism*, 35. For the vitalist interpretation of Shelleyan "life" and its universalism, see Sharon Ruston, *Shelley and Vitality* (London: Palgrave Macmillan, 2009), esp. 2–3, 175–180. Ross Wilson posits a "post-organic animism" in Shelley's poetry that relates the performance of poetry to the production of life and sees *The Triumph of Life* as restaging the familiar "opposition between life and theory." Wilson, *Shelley and the Apprehension of Life* (Cambridge: Cambridge University Press, 2013), 6, 145. For the vitalist concept of suspension in Shelley's poetry and social thought, see Robert Mitchell, *Experimental Life: Vitalism in Romantic Science and Literature* (Baltimore: Johns Hopkins University Press, 2013), 43–73.

110. In Mitchell's impressive analysis, Shelley's formal and philosophical conception of suspension registers "a holding-against that served as the basis for the subsequent development of particular forms of willing" (*Experimental Life*, 63).

111. In an attempt to more accurately account for being within a climate that can no longer abide extraction, Jason Moore has coined the term "capitalocene," which describes the pervasive influence of one economic model on humanity's interaction with nature. See Jason Moore, *Capitalism in the Web of Life: Ecology and the Accumulation of Capital* (London: Verso, 2015), 169–192. In a now infamous essay, Jonathan Kramnick dismantles many of the assumptions behind evolutionary criticism. Kramnick, "Against Literary Darwinism," *Critical Inquiry*

37, no. 2 (2011): 315–347. Goldstein pushes against the trend to denude science studies of any political content in her new book on Romantic materialism (*Sweet Science*, throughout).

112. Mitchell approaches Shelley's conception of the automatic movement of nature/politics in a slightly different way than I do. He contends that the suspension of the movements of the natural/social world inspired by vitalism could be an antidote to the involuntary habits and reactions driven by the hyperanimated modern-media environment that was emerging in Shelley's day (*Experimental Life*, 61–62). According to Mitchell, such a suspension revises Shelley's original political affinity for a dynamic, "progressive materialism" of ceaseless change (62). I argue, however, that Shelley never abandons this progressive materialism—or, by extension, his belief in the mutability of social conditions—but, instead, recognizes that such a movement must be relegated, occasionally, to a background condition of life. Of course, it should be noted that one's ability to *avoid* constant political struggle—to place struggle in the background occasionally—is an argument typically made of representative democracy. In this way, my "automatic radical" Shelley and Mitchell's "passive liberal" Shelley are not that different after all.

113. Shelley, *The Triumph of Life*, 494, 492, 498.

114. As Michael Hardt and Antonio Negri write, the very structure of labor today makes it harder to hide the phenomenon of a shared, common endeavor: "Capitalist abstraction always rests on the common and cannot survive without it," while, "in the age of biopolitical production, the common, which previously was cast as external, is becoming completely 'internalized.'" Hardt and Negri, *Commonwealth* (Cambridge, MA: Harvard University Press, 2009), 159, 283. In *Ecosocialism*, the sociologist Michael Löwy notes that "solar energy, which has never aroused much interest in capitalist societies, . . . must become the object of intensive research and development and play a key role in the building of an alternative energy system." Löwy, *Ecosocialism: A Radical Alternative to Capitalist Catastrophe* (Chicago: Haymarket Books, 2015), 23; and see esp. 1–17. The progressive policy analyst Matt Bruenig posits that democratized investment vehicles such as social wealth funds could provide people with a stable income and act as a transition vehicle toward the communal ownership of production. Bruenig, "Nickel-and-Dime Socialism," Medium, February 2017, https://medium.com/@MattBruenig/nickel-and-dime-socialism-47fcec406295. In an essential chapter from *We Have Never Been Modern*, "Constitution," Bruno Latour reveals the political stakes of an "independent" scientific method in the confrontation between Hobbes and Robert Boyle. Latour, *We Have Never Been Modern*, trans. Catherine Porter (Cambridge, MA: Harvard University Press, 1993), 13–48, esp. 20–21. Finally, Nancy Fraser's recent work has been dedicated to advancing the idea of an economy in which care and emotional labor are equivalent to—or supersede—production and service. Fraser, *Fortunes of Feminism: From State-Managed Capitalism to Neoliberal Crisis* (London: Verso, 2013), esp. 102–122.

115. Shelley, *The Triumph of Life*, 496.

Afterword

ON THE USES OF THE HISTORY OF TECHNOLOGY FOR LITERARY STUDIES AND VICE VERSA

Joseph Drury

The best interdisciplinary scholarship offers something of value to scholars on both sides of the disciplinary divide that it seeks to bridge. My contribution to this collection is to try to use the preceding essays in this volume to draw some general observations about what literary historians and historians of technology can learn from each other when they align their objects of study. What is distinctive about the practice of reading literature through and in relation to the history of technology? What do the archives and methods of the history of technology reveal about literary texts and literary history that other approaches do not? And conversely, what can literature and the kinds of textual analysis practiced by literary scholars in this volume contribute to the history of technology that other sources and methods cannot? The opportunity to reflect on these questions is rarer than one might think. Thanks in part to the encouragement of academic societies and their flagship journals, scholars of literature and science have regularly paused to map their field and reflect on emerging problems and methods.[1] But the field of literature and technology lacks these institutional supports, and its scholars as a result tend to stay focused on particulars. Those who do this kind of work—if they think of themselves as belonging to such a field at all—are left hoping to find their concerns addressed in passing in essays about literature and science. But while the making of scientific knowledge almost always requires some kind of technology, technologies do not always involve science or scientists. Technologies are ways of doing things, not just ways of knowing. As such, they extend deep into the rhythms of everyday life in a way that is less often the case with scientific knowledge. The key questions and methodological problems raised by reading literature in relation to the history of technology are therefore distinct from—and need to be theorized independently from—those raised by studying it in relation to the history of science.

Let me start by addressing the first pair of questions—the ones about the value of the history of technology for literary studies—since they are the ones with which my own work has been most concerned.[2] The most important general observation suggested by these essays, to my mind, is that the history of technology offers literary scholars valuable new resources and an expanded vocabulary for thinking about the *instrumental* dimensions of literary forms—that is, the uses and effects of different genres, techniques, styles, and the like, and the various means by which literary texts mediate the subjectivity and behavior of authors and readers. One way of thinking about the history of literary studies is as a series of interdisciplinary experiments in the ontology of the literary object. In turning to the interpretive methods and resources of other disciplines, literary scholars have repeatedly—and productively—made the text resemble the objects of study in those disciplines. Critics who studied Derrida viewed the text as an unstable semiotic system; those influenced by Freud or Lacan saw it as a kind of dream-work or neurotic symptom; scholars who drew on Marx and Foucault uncovered ideology and technologies of power; and so on. That is not to say that there has ever been a *purely* literary approach to a literary text; even methods presented as such—New Criticism or structural narratology, for example—in fact made use of other disciplines (psychology, linguistics, anthropology) to insist on the autonomy of the literary object.[3] One might even say that all literary scholarship is to some extent interdisciplinary; work that appears not to be so is merely drawing on other disciplines unconsciously or in ways that have become so naturalized to literary study as to be imperceptible. It makes sense, then, that in turning to the history of technology, several essays in this collection should treat the literary object—whether a genre, text, or formal device—as a kind of technology, a means by which some end is achieved. Laura Francis argues that John Webster's *The Duchess of Malfi* uses the baroque theater as a laboratory or experimental "instrument" for testing the epistemic limits of contemporary optical technologies and natural philosophy. In Thomas Oldham's reading of the Scriblerian farce *Three Hours after Marriage*, he notes how the potentially emancipatory implications of the play's parody of the virginity-test scene from another Jacobean tragedy, Thomas Middleton and William Rowley's *The Changeling*, are undermined by the specular economies of theatrical representation, which, like the supposedly enlightened gynecological treatise to which Oldham compares it, functions as an "instrument" for turning women's bodies into objects of male consumption and control. Likewise, Jamison Kantor explores Percy Shelley's fascination with the technological marvels of the early nineteenth century, observing that the "mechanical regularity" of the verbal refrains that Shelley inserted into his protest poem *The Mask of Anarchy* makes them resemble "mnemonic devices" that "automate thought for readers," while his use of allegory to lend meaning to contingent historical events operates as "an engine that regiments the pieces and narrative of public affairs." In these essays, the exploration of a text's intersections with the history of technology produces a special attentiveness to the different *functions* of various literary techniques, formal devices, and modes of representation.

A particular advantage of studying the entanglements of literature and technology in the period covered by this volume is that the nature and extent of the affinities between them was the subject of extensive legal debate. In limiting the property rights of authors to a period of fourteen years, the legislators responsible for the 1710 Statute of Anne, which has been called "the world's first copyright statute," drew on the system of patent law that had long protected technological innovations.[4] As a result, one of the key issues in the half century of legal debates prompted by the statute was how, if at all, literary inventions differed from mechanical ones. In an opinion written for *Donaldson v. Becket*, the 1774 House of Lords ruling that settled the debate decisively in favor of statutory limits by equating literary with technological innovation, Baron James Eyre claimed that literary works and mechanical inventions were both forms of intellectual labor that involved the ordering of different parts into a whole in order to achieve some purpose: "In a mechanic invention the Corporeation of Parts, the Junction of Powers, tend to produce some one End. A literary Composition is an Assemblage of Ideas so judiciously arranged, as to enforce some one Truth, lay open some one Discovery, or exhibit some one Species of mental Improvement. A mechanic Invention, and a literary Composition, exactly agree in Point of Similarity."[5] As Zachary Mann argues in his contribution to this collection, taking such claims seriously requires incorporating the history of technology into literary history in a way that "reconceptualizes eighteenth-century knowledge production in industrial terms." The relations between author, text, and reader in the Enlightenment period come to seem analogous to those between "philosopher-inventors," machines, and their users. Creativity, composition, thinking itself all begin to be defined through a dualistic, hierarchical opposition to forms of intellectual labor—imitation, quotation, compilation, indexing—long thought to belong to a broad continuum of humanistic literary practices but now deemed merely "mechanical," alienated, even "subhuman." The most important sign of an author's genius becomes the capacity for "invention" and the arrangement of ideas conceived in Lockean terms as discrete units, while the value of a literary object—whether a form, a genre, or individual text—consists in its utility to the reader.

This reconceptualization of literary labor might seem reductive, but as this collection makes clear, there is in fact a great deal to be said about the uses of literary objects. Indeed, sustained attention to the uses of literature ought to be especially welcome now, given recent complaints about the impoverished state of contemporary critical discourse on the ways that readers engage with texts. Joshua Landy, for example, bemoans his students' narrow preoccupation with the "propositional content" of works of fiction, which he blames on a generation of English professors and teachers who have "systematically—albeit unwittingly—engaged on a long-term campaign of misinformation, relentlessly persuading would-be readers that fictions are designed to give them useful advice."[6] All but forgotten, he thinks, are the many other designs that fictions have always had on us, such as strengthening our capacity for empathy or moral action, enhancing our knowledge of the world, or refining our ability to give shape and meaning to life experience. Rita Felski has echoed these concerns in her recent work, noting that "the current canon of theory yields a paucity of rationales for attending to literary objects," though she attributes this predicament to the dominance of two opposing but equally misguided critical practices: a "theological" approach that insists on the alterity, ineffability, and thus "impotence" of literary works and an "ideological" one that whittles the notion of a text's use down "to the bare bones of political and ideological function." As a solution, Felski turns to the phenomenological tradition to call for renewed engagement with what she terms the "worldly aspects of literature"—that is, the variety and complexity of its real-world uses and effects. Such an engagement need not entail neglect of its aesthetic qualities, she argues; on the contrary, "to propose that the meaning of literature lies in its use is to open up for investigation a vast terrain of practices, expectations, emotions, hopes, dreams, and interpretations."[7]

Some notable recent attempts to explore this terrain have hinted at a key role for science and technology studies. Felski herself has called for new "postcritical" methods of reading that eschew the hermeneutics of suspicion in favor of an approach informed by Bruno Latour's actor-network theory. Instead of assuming a position of detachment from the text, the postcritical reader presumes the "coimplication and entanglement of text and critic"; and rather than treating texts as inert objects of analysis incapable of transcending the historical context in which they were composed, the postcritical reader treats them the way Latour treats "nonhuman actors" in a network—not as "passive intermediaries" but as "active mediators," "coactors that make things happen" across time and space.[8] Yet, despite making a powerful case for actor-network theory as a resource for literary scholars, Felski ignores the importance of science and technology studies to its origins and practice and as a result fails to recognize its potential value to the postcritical project for which she advocates. Latour himself recalls that he developed actor-network theory due to his dissatisfaction with the so-called Strong Programme within the sociology of scientific knowledge, which took the then-fashionable position that the "hard facts of science" were socially constructed all the way down. The method that he and his collaborators developed as an alternative transformed not just the sociology of knowledge but social science as a whole by insisting that nothing is *merely* socially constructed and that all human activities involve networked entanglements of the social and the material, subjects, and objects. This origin story—or "*felix culpa*," as he calls it—explains why Latour thinks that one of the best ways for social scientists to make visible the agency of nonhuman actors remains ethnographic studies of technology. In "the artisan's workshop, the engineer's design department," or "the scientist's laboratory," he argues, objects "live a clearly multiple and complex life," observable not just in the various "meetings, plans, sketches, regulations, and trials" that lead to the establishment of the network in which they are an actor—and after which they disappear from view—but also in their interaction with existing social agencies.[9]

Some recent reflections on literary form have suggested that the study of technology might prove as valuable a resource for exploring the agency of literary objects as it has been for other nonhuman actors. For Henry Turner, Latour's understanding of form as the "principle of translation" responsible for forging associations in a network offers literary scholars a way to move beyond criticism's "reductive" preoccupation with what forms mean or represent toward a new focus on their function: "form *does* things, it doesn't simply mean things," he insists. But reimagining literary forms along these lines, he argues, will require critics to break with some long-standing habits: "one of the most pressing tasks confronting literary criticism today will be to expand concepts of form beyond linguistic and textual models" to include "*all* meaningful artifacts, of whatever type—written texts, yes, but also physical specimens, instruments readings, photographs, models of all kinds," since all of these objects "possess at least a minimal structural complexity and use signifying elements in ways that are conventional and thus formalized."[10] Taking up this challenge, Caroline Levine has drawn on actor-network theory to explore the way forms connect "things, people, and ideas to each other," introducing the concept of "affordance" from design theory to think through "the particular constraints and possibilities that different forms afford." Just as a material such as glass affords "transparency and brittleness," for example, and a simple device such as a doorknob affords "turning, pushing, and pulling," so a literary form such as rhyme affords "repetition, anticipation, and memorization" and "the triple-decker novel affords elaborate processes of character development in multiplot social contexts."[11] Levine ultimately turns out to be more interested in social than technical systems, but her basic proposition that design theory offers a framework for thinking about the agency of forms in general is a suggestive one for anyone interested in the intersections of literature and technology.

This volume explores some of the ways that scholars might use the history of technology to think about the agency of literary objects. For those seeking to understand the agency of nonhuman actors in general, Latour argues, the next best thing to studying technologies as they are being created is to bring them back into view "using archives, documents, memoirs, museum collections, etc., to artificially produce, through historians' accounts, the state of crisis in which machines, devices, and implements were born."[12] If design theory provides a theoretical framework for the symmetrical analysis of literary and material forms, the history of technology extends that framework to include the human and nonhuman actors common to the different networks in which they operate. Thus, Emily West proposes that the tangled provenance, composition, and social function of the gothic lantern that Horace Walpole installed above the staircase at Strawberry Hill makes it an ideal interpretive framework for examining the design of *The Castle of Otranto* (1764), whose depiction of perverse genealogical lines is structured by the same "aesthetics of queer time" that the lantern materializes. Meanwhile, Kevin MacDonnell finds in the serpentine line traced by the linkage in James Watt's steam engine as it mediates the divergent motions of the piston and beam the same Hogarthian "principles of design" that inspired the simultaneously progressive and digressive narrative machinery of Laurence Sterne's *Tristram Shandy*. To be sure, such explorations of the shared cultural contexts shaping a technology and a literary form are not entirely new to eighteenth-century scholarship. John Bender pioneered this kind of work in the 1980s with his classic study of the overlapping discursive practices and procedures of the emerging penitentiary system and the realist novel.[13] But in Bender's work, as in other New Historicist studies, form is understood to determine function. The assumption is that because the novel works like a prison, it must always have the same disciplinary function as one. Formal similarities may *sometimes* indicate functional ones; indeed, it certainly makes sense that the atmospheric effects and ideological subversions accomplished by Walpole's gothic lantern should be replicated in his gothic fiction. But one of the key insights of modern science and technology studies reflected in these essays is that the function of a technical artifact depends on the particular circumstances of its use. To assume otherwise, argues the philosopher Andrew Feenberg, is to be guilty of a kind of fetishism: "like price, function is a relational term which we attribute to the object as a real quality." In reality, however, "the function of any technology is relative to the organizations that create and control it and assign it a purpose."[14] Thus, although both Watt's linkage and Sterne's narrative, according to MacDonnell, "negotiate the systematizing nature of design with the elastic quality of empirical reality," they do so to very different ends. Unlike Watt's steam engine, which revolutionized industrial production and, as MacDonnell notes, helped launch the Anthropocene, Sterne's narrative is an "antimachine" that undertakes a similar form of mediation but for the purposes of "Enlightenment self-critique." In both these essays, the history of technology is deployed not, as one might expect, to assimilate these fictions' notorious formal eccentricities to normative Enlightenment productive and reproductive systems but to explore the aesthetic principles and cultural contexts informing their resistance to them.

But what of the final question I began with, the one about what literature and literary analysis has to offer the history of technology? The reflection on the uses of literary objects in these essays ought to be as valuable to historians of technology as it is to literary historians. Urging historians of science to pay attention to recent debates in literary studies, Henry Turner argues that "literary critics and historians of science alike should be undertaking an analysis of the many *networks of forms* that constitute our objects of inquiry and that allow any mode of communication, thought, or collective association to take place."[15] Historians of technology have just as much to gain, since literary forms play an equally critical role in the invention and evolution of technologies. Although Steven Shapin and Simon Schaffer were primarily concerned with providing an account of the origins of the scientific method, the attention they paid to the "literary technology" of the experimental narrative in *Leviathan and the Air-Pump* also revealed a good deal about the textual forms that helped establish the epistemological authority—and thus utility and influence—of modern scientific instruments.[16] Recent literary studies of the balloon craze of the 1780s suggest that the analysis of literary forms can yield similar dividends in areas that are more traditionally of interest to historians of technology. John Robbins, for example, has noted that Elizabeth Inchbald's immensely popular 1784 farce *The Mogul Tale*, about a group of English balloonists who are blown off course and end up in the harem of an Indian prince, appeared before any manned ballooned flights had taken place in Britain. The play thus functioned as "a kind of ballooning demonstration in its own right," one of a number of theatrical performances in the period that sought to turn the theater into a "tool for shaping opinions about the new technology of ballooning, and technology in general, to the English public at large."[17] This volume features other examples of literary forms that were used to influence public opinion about technological innovation, from the revenge tragedy that John Webster turned into a means of scrutinizing the truth-making powers of the new philosophy and its technologies to the didactic "popular tale" that Maria Edgeworth adopted to "neutralize the disruptive potential" of the optical telegraph at the turn of the nineteenth century. In my own work, I have also pointed to the power of the "it-narrative" to shape attitudes to disruptive new transportation technologies such as the post-chaise, which helped double the speed at which passengers could travel when it was first introduced to Britain in the 1750s.[18] Historians of technology will not need to be reminded of the central role that public opinion plays in sustaining the networks involved in the creation and use of new technologies or that successful technological innovation requires much more than the solitary labor of heroic inventors. In the United States, one need only think of the chilling effect that the *Challenger* disaster had on space exploration or the influence of climate-change denialism on the stalled transition to electric and hybrid vehicles to recognize the constitutive role public attitudes and imaginaries continue to play in technological development.

The "cultural turn" taken by the history of technology over the past two decades points to other ways that historians of technology might make use of the kind of literary scholarship found in this volume. As Carroll Pursell has observed, today's historians of technology no longer confine themselves to causal explanations of how and why certain technologies came to be invented, nor are the questions they ask of a tool or machine limited to ones about where it comes from, how it works, or what it does. Influenced by cultural studies, they now seek to *interpret* technologies, viewing a machine as "both a cultural practice and cultural production" about which one might also ask, "what does it mean?"[19] The cross-disciplinary origins of American studies have meant that U.S. literary history, starting with Leo Marx's seminal *The Machine in the Garden*, has been a particularly fruitful field for work on the cultural meanings of technologies.[20] But even half a glance at the book lists of the major academic publishers will show that the past two decades have seen a growing number of cultural histories of technology written by scholars of British literature.[21] Literary scholars in these studies use their expertise in textual analysis to generate arguments about the shifting cultural meanings of technologies such as the windmill and the mechanical clock, the railway and the radio, the automobile and the internet.

Several essays in this volume illustrate the advantages of this approach. Thinking about the meaning of a technology requires looking beyond the circumstances of its invention to its use. To support the argument that the French optical telegraph came "to stand for more than the device itself" in British culture, becoming in the course of the 1790s a potent symbol of French military power and the threat of Jacobinism, Deven Parker focuses not on its inventor, Claude Chappe, though his nationality is obviously relevant, but on media reports and images of its real and potential uses. In British newspaper accounts, she observes, the French military's use of the telegraph to transmit news of victories back to the government in Paris were conflated with the victories themselves, thereby suggesting "a causal link" between the two, while James Gillray's eerie image of Charles James Fox as a human telegraph lighting the invading French fleet's way to London figures the device as an "extension" of the revolutionary French government that was threatening "to infiltrate Britain both physically and ideologically." Indeed, Parker implies that arguments about the priority of a particular invention—a staple of traditional history of technology—are of interest not because they explain anything about the evolution of a technology but for what they reveal about its cultural meaning at the time such arguments are made. The "historical revisionism" motivating the flood of publications claiming to prove that the British had invented the telegraph first merely betrays national anxieties about the undeniable superiority of the French when it came to making use of it. Parker's analysis also shows that the shift in focus from innovation to use that cultural history requires, while a worthy historiographic enterprise in its own right, can also make for better, more robust causal explanations of a technology's evolution: "invention and innovation rarely lead to use," the historian David Edgerton observes, "but use often leads to invention and innovation."[22] As Parker makes clear, the various British inventors of optical telegraphs had much less of an impact on the development of a telegraph system in Britain than did the terrifying media reports of its successful use by the French.

Attending to the uses and meanings of technologies rather than innovations also makes it possible to create space for a wider range of experiences and voices. When told in terms of innovation and production, the history of technology typically becomes "a story of the deeds of great men."[23] Cultural histories and use-centered accounts, on the other hand, "shift attention from the new to the old, the big to the small, the spectacular to the mundane, the masculine to the feminine, the rich to the poor."[24] The novelist Maria Edgeworth moves into the center of a narrative once dominated by her father, the inventor and Lunar Society member Richard Lovell Edgeworth. The hinge receives as much attention as Watt's steam engine, the colonial or "creole" appropriations of a technology as much as its design and construction at the imperial center. Just as importantly, such an approach also makes it possible to consider technologies that were valuable and meaningful not because of their novelty but because of their association with the past or because of their disruption of the kind of teleological narratives characteristic of histories of technological innovation. As Emily West notes in her chapter in this collection, the material composition of Walpole's gothic lantern "revises the way that technological objects were called on across the eighteenth century to manifest Enlightenment modernity as a progression from the superseded past." Such objects remind us not to conflate the history of technology with the history of innovation and that, as Edgerton observes, "technologies do not only appear, they also disappear and reappear, and mix and match across the centuries."[25]

Of course, the uses to which technologies are put in the imaginative works that literary scholars typically focus on are fictional, sometimes even fantastical, which means the evidence they provide is of a distinctive kind. In his classic essay "Time, Work-Discipline and Industrial Capitalism," E. P. Thompson used the detailed descriptions of agricultural labor in the georgic poetry of the working-class poets Stephen Duck and Mary Collier as evidence of the way new clock-regulated work regimens began to change people's experience of time in the eighteenth century. But when he alluded to the famous opening scene from *Tristram Shandy*, it was not of course to suggest that any real people might have used clocks to regulate their sexual activity the way Walter Shandy does. Jokes are evidence not of realities but of attitudes, or what Raymond Williams would later call the "structures of feeling" of a particular historical moment. The opening scene of *Tristram Shandy*, like several other episodes in Sterne's novel, comments trenchantly on the "intimate levels" of human experience clocks and watches were beginning to reach by the mid-eighteenth century.[26] It is precisely this insight into the "subjective experience" of a machine or device, however, that Rosalind Williams argues is missing from much history of technology, which even in its most methodologically sophisticated forms typically focuses on the design, construction, and operation of objects. Although social constructionist studies of technology bring users, consumers, and maintainers into the foreground, they are more interested in "behavior, especially aggregate behavior," than subjective experience, she notes, while actor-network theory prides itself on "putting conscious and non-conscious agents on the same level." Fictions such as *Tristram Shandy*, on the other hand—Williams's examples are Robert Louis Stevenson's South Pacific narratives—demonstrate "the value of imaginative literature both as a record of, and even more as a source of insight into, historical experience." Although there are other ways that historians can access the phenomenological dimensions of a technology, Williams argues that literature is an especially useful resource because it "approaches the world simultaneously as externally experienced—a field out there—and as internally experienced—inward feelings, passions and thoughts."[27] In this volume, therefore, Erik Johnson uses readings of fiction by Daniel Defoe and Mary Hearne to challenge common assumptions about the effect of new mechanical devices on the experience of time in the eighteenth century. Rather than illustrating how clocks and watches enabled individuals to order their experience by allowing them to map their ideas in time and space, as Locke had suggested, the disruptions and short-circuits in the detailed "temporal accountings" of Robinson Crusoe on his island and Hearne's beleaguered heroine reveal the "fragility" of attempts to impose this kind of order, betraying "a deep suspicion of the modern, technologically mediated experience of time."

One final way that this volume asks historians to think about literature is as an archive for studying the *imagined* uses of technologies. Edgeworth's "Lame Jervas" is a record not of the way the optical telegraph was actually used—Parker looks to newspapers for that—but of the uses that were imagined for it during the period when interest in its development was at its height. Tippoo Sultan's exultant response to Jervas's demonstration of his system articulates British fears about the dangers of the telegraph falling into the hands of foreign despots and British fantasies of technological and cultural supremacy. The virginity test that Dr. Fossile gives his bride in *Three Hours after Marriage* is ludicrous, but its imagined uselessness, as Oldham demonstrates in his essay, reflects real changes in medical science brought about by the introduction of the forceps. Indeed, as Simone Müller and Heidi Tworek have recently argued, the fictional uses of a technology have often proved as important for understanding its historical meaning and development as the actual ones: "imagined uses shaped cultural narratives about technology as harbingers of modernity, as agents of universal peace, or, in the case of distorted man-machine relationships, as the end of mankind," often structuring the political framing of research and use in ways that influence "the physical development of technologies." As an example, they point to the impact of science fiction by Arthur C. Clarke and Robert Heinlein on cryonics, the practice of freezing the dead for future reanimation. But one could point to earlier instances from the seventeenth and eighteenth centuries. If we are to believe *The Clockmakers Outcry against the Author of the Life and Opinions of Tristram Shandy* (1760)—though unlike E. P. Thompson, I suspect we are not—the use that Walter Shandy made of his house clock had a material effect on the trade of English clockmakers in the months following the publication of the work's first two volumes. Literary representations of the Aeolian harp had a particularly direct and palpable impact on its popularity in the second half of the eighteenth century. The Scottish musician and publisher James Oswald is said to have been inspired to design the instrument in the 1740s after reading a description of music being produced by the wind in Homer, while sales of the instrument from his shop in London were given an early boost by its appearance in literary works by fellow Scots, *The Castle of Indolence* (1748) by the poet James Thomson and Tobias Smollett's picaresque novel *The Adventures of Ferdinand Count Fathom* (1753). But Müller and Tworek point to two other advantages of incorporating imagined uses into the history of technology. First, extending the concept of use to include fictional or poetic representations allows historians to broaden the social groups involved in technological innovation and appropriation by including those who may never have had any contact with a technology but were nonetheless powerfully affected by it. As an example, they point to the effect of the electrical telegraph on the imagination of the Victorian public: people who could not afford to send a telegram nonetheless read news sent by it and were inspired to imagine its global social effects: "without necessarily ever physically using the technology, these actors still shaped the financial, political, or marketing regimes that provided the structural conditions of technological development." Second, attending to imagined uses makes space for histories of "imagined futures"; whether or not literary visions of technological futures become a reality, they have "fundamentally shaped the present" by affecting how innovators thought and planned.[28] The real Tipu Sultan did not use an optical telegraph, while Richard Lovell Edgeworth's design was never implemented in England and only briefly in Ireland. But it is not hard to see how Maria Edgeworth's account of the use that Tipu *might* have made of it would have influenced the development of the optical telegraph systems that were actually constructed in Britain and parts of the empire during the first few decades of the nineteenth century.

Let me close with a final observation about the value of this kind of work today. In a recent essay on the potential "postdisciplinary liaisons" that might be forged between science studies and the humanities, Mario Biagioli notes that, unlike many other academic disciplines that emerged after World War II, science studies has not been threatened by the rise of cultural or postcolonial studies, or by the changes in U.S. immigration patterns that affected the study of European literature and culture, or even by the shifts in funding and research priorities that followed the end of the Cold War. One explanation for this, he argues, is that science studies—including the study of technology—has become ever more relevant to the world in which we live: "if around 1950 the popular imaginary placed science close to the military and away from the home, today's technoscience frames our everyday life at all levels, down to our notion of the self."[29] As I write this in the summer of 2020, the planet is in the grip of a pandemic that has caused, so far, the death of more than half a million people worldwide and the disability of hundreds of thousands, perhaps millions more; it has destroyed livelihoods, put millions of people out of work, shuttered shops, restaurants, offices, schools, universities, museums, galleries, libraries, cinemas, and theaters; it has changed the way people work, travel, and socialize; the way we teach, learn, and study; the way we educate our children and the way we vote; the way we fall in love and the way we die, mourn, and grieve. We spend our days reading and watching news reports about the latest vaccine trial or epidemiological study published in *Nature* or the *Lancet*. On social media, we practice armchair immunology and harangue each other about public-health measures such as wearing face masks and social distancing. To make sense of all these changes, we have turned sometimes to the history of science, medicine, and technology, to stories about the invention of the smallpox vaccine, the 1918 flu pandemic, and the public infrastructure projects launched in the nineteenth century to prevent cholera outbreaks. But we have also sought solace in works of fiction, from classics like Daniel Defoe's *Journal of the Plague Year* and Albert Camus's *The Plague* to contemporary postapocalyptic novels by Colson Whitehead, Margaret Atwood, and Emily St. John Mandel—material that does not merely help explain how we arrived in this situation or how previous generations were able to escape similar ones but that also offers insight into the meaning of our experience by giving context and coherence to our daily feelings of fear, anger, or despair. In such a world, the cross-disciplinary study of every aspect of the technologies that have shaped and continue to shape human life on this planet seems more essential than ever.

NOTES

1. For some recent examples, see the essays under the heading "History of Science and Literature and Science: Convergences and Divergences" in the "Focus" section of the journal *Isis* 101, no. 3 (2010); Sabine Sielke, "Science Studies and Literature," *Anglia* 133, no. 1 (2015): 9–21; and the special issues of the *Journal of Literature and Science* (10, no. 1, 2017) and *Configurations* (26, no. 3, 2018). For a recent survey of the field focusing on the long eighteenth century, see

my own article, "Literature and Science in Enlightenment Britain: New Directions," *Literature Compass* 14, no. 6 (2017), https://compass.onlinelibrary.wiley.com/doi/10.1111/lic3.12395.

2. See Joseph Drury, *Novel Machines: Technology and Narrative Form in Enlightenment Britain* (Oxford: Oxford University Press, 2017).

3. On the influence of behaviorist psychology, for example, on early New Criticism, see Joshua Gang, "Behaviorism and the Beginnings of Close Reading," *ELH* 78, no. 1 (2011): 1–25. For a good recent account of the importance of Saussurean linguistics and Claude Lévi-Strauss's anthropology to structuralist narrative theory, see Kent Puckett, *Narrative Theory: A Critical Introduction* (Cambridge: Cambridge University Press, 2016), 205–214, 215–222.

4. Mark Rose, *Authors and Owners: The Invention of Copyright* (Cambridge, MA: Harvard University Press, 1993), 5.

5. Quoted in ibid., 88.

6. Joshua Landy, *How to Do Things with Fictions* (Stanford, CA: Stanford University Press, 2012), 3.

7. Rita Felski, *The Uses of Literature* (Oxford: Blackwell, 2008), 2, 5, 7–8.

8. Rita Felski, *The Limits of Critique* (Chicago: University of Chicago Press, 2015), 152, 164, 180.

9. Bruno Latour, *Reassembling the Social: An Introduction to Actor-Network Theory* (Oxford: Oxford University Press, 2005), 94, 80.

10. Henry S. Turner, "Lessons from Literature for the Historian of Science (and Vice Versa): Reflections on 'Form,'" *Isis* 101, no. 3 (2010): 586, 582. For Latour's reflections on the function of forms in networks and his definition of a form as "something which allows something else to be transported from one site to another," see Latour, *Reassembling the Social*, 222–223.

11. Caroline Levine, *Forms: Whole, Rhythm, Hierarchy, Network* (Princeton, NJ: Princeton University Press, 2015), 153n22, 6. Felski also discusses the potential of this concept for literary studies in *The Limits of Critique*, 164–165. The term was coined by the psychologist James J. Gibson to explain the interaction of animals with their environment, but it was popularized by the cognitive scientist Don Norman in *The Psychology of Everyday Things* (New York: Basic Books, 1988).

12. Latour, *Reassembling the Social*, 81.

13. See John Bender, *Imagining the Penitentiary: Fiction and the Architecture of Mind in Eighteenth-Century Britain* (Chicago: University of Chicago Press, 1987).

14. Andrew Feenberg, *Questioning Technology* (London: Routledge, 1999), 117.

15. Turner, "Lessons from Literature," 586.

16. Steven Shapin and Simon Schaffer, *Leviathan and the Air-Pump: Hobbes, Boyle, and the Experimental Life* (Princeton, NJ: Princeton University Press, 1985), 25.

17. John Robbins, "Up in the Air: Balloonomania and Scientific Performance," *Eighteenth-Century Studies* 48, no. 4 (2015): 531, 528.

18. Drury, *Novel Machines*, 125–126.

19. Carroll W. Pursell, "Technologies as Cultural Practice and Production," *Technology and Culture* 51, no. 3 (2010): 721, 716.

20. Leo Marx, *The Machine in the Garden: Technology and the Pastoral Ideal in America* (Oxford: Oxford University Press, 1964).

21. See, for example, Tim Armstrong, *Modernism, Technology, and the Body: A Cultural Study* (Cambridge: Cambridge University, 1998); Pamela Thurschwell, *Literature, Technology, and Magical Thinking, 1880–1920* (Cambridge: Cambridge University Press, 2001); Jay Clayton, *Charles Dickens in Cyberspace: The Afterlife of the Nineteenth Century in Postmodern Culture* (Oxford: Oxford University Press, 2003); Nicholas Daly, *Literature, Technology, and Modernity, 1860–2000* (Cambridge: Cambridge University Press, 2004); Jonathan Sawday, *Engines of the Imagination: Renaissance Culture and the Rise of the Machine* (London: Routledge, 2007); Jessica Wolfe, *Humanism, Machinery, and Renaissance Literature* (Cambridge: Cambridge University Press, 2004); Enda Duffy, *The Speed Handbook: Velocity, Pleasure, Modernism* (Durham, NC: Duke University Press, 2009); Laura Otis, *Networking: Communicating with Bodies and Machines in the Nineteenth Century* (Ann Arbor: University of Michigan Press, 2011); Tamara Ketabgian, *The Lives of Machines: The Industrial Imaginary in Victorian*

Literature and Culture (Ann Arbor: University of Michigan Press, 2016); Rajani Sudan, *The Alchemy of Empire: Abject Materials and the Technologies of Colonialism* (New York: Fordham University Press, 2016); Andrew Gaedtke, *Modernism and the Machinery of Madness: Psychosis, Technology, and Narrative Worlds* (Cambridge: Cambridge University Press, 2017); Clare Brant, *Balloon Madness: Flights of the Imagination, 1783–1786* (Woodbridge, UK: Boydell Press, 2017).

22. David Edgerton, "From Innovation to Use: Ten Eclectic Theses on the Historiography of Technology," *History and Technology* 16, no. 2 (1999): 121.

23. Carroll Pursell, "Seeing the Invisible: New Perceptions in the History of Technology," *Icon* 1 (1995): 11.

24. David Edgerton, *The Shock of the Old: Technology and Global History since 1900* (London: Penguin, 2019), xxiv.

25. Ibid., xxii.

26. E. P. Thompson, "Time, Work-Discipline, and Industrial Capitalism," *Past & Present* 38 (1967): 63, 73, 57. Paul Glennie and Nigel Thrift in *Shaping the Day: A History of Timekeeping in England and Wales, 1300–1800* (Oxford: Oxford University Press, 2009) challenge many of Thompson's conclusions, but they also use literary works as evidence, citing the poetry of William Langland and John Clare and engaging thoughtfully with the literary scholarship of Mikhail Bakhtin and Stuart Sherman, among others. For Raymond Williams's most extended discussion of the concept of "structures of feeling," see *Marxism and Literature* (Oxford: Oxford University Press, 1977), 128–135.

27. Rosalind Williams, "The Lantern-Bearers of the History of Technology," *History and Technology* 29, no. 3 (2013): 262, 271.

28. Simone M. Müller and Heidi J. S. Tworek, "Imagined Use as a Category of Analysis: New Approaches to the History of Technology," *History and Technology* 32, no. 2 (2016): 106–108.

29. Mario Biagioli, "Postdisciplinary Liaisons: Science Studies and the Humanities," *Critical Inquiry* 35 (2009): 818.

Bibliography

Alker, Sharon. "Explosive Potential: Radicals and Chemistry in Maria Edgeworth's Fiction." *European Romantic Review* 22, no. 1 (2011): 1–18.

Alkon, Paul K. *Defoe and Fictional Time*. Athens: University of Georgia Press, 1979.

Allan, D.C.G. *William Shipley, Founder of the Royal Society of Arts: A Biography with Documents*. Corr. ed. London: Scolar Press, 1979.

Allen, Robert C. "The Industrial Revolution in Miniature: The Spinning Jenny in Britain, France, and India." *Journal of Economic History* 69, no. 4 (2009): 901–927.

Althusser, Louis. *Lenin and Philosophy and Other Essays*. New York: Monthly Review Press, 2001.

Amster, Mara. "Frances Howard and Middleton and Rowley's *The Changeling*: Trials, Tests, and the Legibility of the Virgin Body." In *The Single Woman in Medieval and Early Modern England: Her Life and Representation*, edited by Laurel Amtower and Dorothea Kehler, 211–232. Tempe: Arizona Center for Medieval and Renaissance Studies, 2003.

Anderson, Susan. *Echo and Meaning on Early Modern English Stages*. New York: Palgrave Macmillan, 2017.

Arago, François. *Historical Eloge of James Watt*. Translated by James Patrick Muirhead. London: John Murray, 1839.

Aristotele's Master-Piece. London, 1684.

Armintor, Deborah Needleman. "The Sexual Politics of Microscopy in Brobdingnag." *SEL: Studies in English Literature, 1500–1900* 47, no. 3 (2007): 619–640.

Armstrong, Nancy. "The Other Side of Modern Individualism: Locke and Defoe." In *Individualism: The Cultural Logic of Modernity*, edited by Zubin Meer, 111–119. Lanham, MD: Lexington Books, 2011.

Armstrong, Tim. *Modernism, Technology, and the Body: A Cultural Study*. Cambridge: Cambridge University Press, 1998.

Augustine. *Confessions*. Translated by Carolyn J. B. Hammond. 2 vols. Loeb Classical Library 26–27. Cambridge, MA: Harvard University Press, 2014–2016.

Austen, Jane. *Northanger Abbey*. In *Northanger Abbey and Other Stories*, edited by James Kinsley, John Davie, and Claudia L. Johnson, 3–187. Oxford: Oxford University Press, 2003.

Austin, J. L. *How to Do Things with Words*. Cambridge, MA: Harvard University Press, 1962.

Bacon, Francis. *The New Organon*. Translated and edited by Lisa Jardine and Michael Silverthorne. Cambridge: Cambridge University Press, 2000.

Badiou, Alain. *Being and Event*. Translated by Oliver Feltham. New York: Continuum, 2007.

Baine, Rodney M. "*Roxana*'s Georgian Setting." *SEL: Studies in English Literature, 1500–1900* 15, no. 3 (1975): 459–471.

Baines, Edward. *History of the Cotton Manufacture in Great Britain*. London: H. Fisher, R. Fisher, and P. Jackson, 1835.

Baines, Paul, and Pat Rogers. *Edmund Curll, Bookseller*. Oxford: Clarendon Press, 2007.

Balizet, Ariane M. "'Drowned in Blood': Honor, Bloodline, and Domestic Ideology in *The Duchess of Malfi* and *El Médico De Su Honra*." *Comparative Literature Studies* 49, no. 1 (2012): 23–49.

Baridon, Michael. "Hogarth's 'Living Machines of Nature' and the Theorisation of Aesthetics." In *Hogarth: Representing Nature's Machines*, edited by David Bindman, Frédéric Ogée, and Peter Wagner, 85–101. Manchester: Manchester University Press, 2001.

Barret, Robert. *A Companion for Midwives, Child-Bearing Women, and Nurses*. London, 1699.

Baucom, Ian. *Specters of the Atlantic: Finance Capital, Slavery, and the Philosophy of History*. Durham, NC: Duke University Press, 2005.

Beck, Lewis White. "World Enough, and Time." In *Probability, Time, and Space in Eighteenth-Century Literature*, edited by Paula Backscheider, 113–139. New York: AMS Press, 1979.

Behrendt, Stephen. *Shelley and His Audiences*. Lincoln: University of Nebraska Press, 1989.

Bellanca, Mary Ellen. "Science, Animal Sympathy, and Anna Barbauld's 'The Mouse's Petition.'" *Eighteenth-Century Studies* 37, no. 1 (2003): 47–67.

Bender, John. *Imagining the Penitentiary: Fiction and the Architecture of Mind in Eighteenth-Century Britain*. Chicago: University of Chicago Press, 1987.

Benjamin, Walter. *The Origin of German Tragic Drama*. Translated by John Osborne. Brooklyn. New York: Radical Thinkers, 2009.

Bentham, Jeremy. *Plan of Parliamentary Reform*. London: R. Hunter, 1817.

Berry, Ralph. *The Art of John Webster*. Oxford: Clarendon Press, 1972.

Biagioli, Mario. "Postdisciplinary Liaisons: Science Studies and the Humanities." *Critical Inquiry* 35 (2009): 816–833.

Bicks, Caroline. *Midwiving Subjects in Shakespeare's England*. Aldershot, UK: Ashgate, 2003.

———. "Stones like Women's Paps: Revising Gender in Jane Sharp's Midwives Book." *Journal for Early Modern Cultural Studies* 7, no. 2 (Fall–Winter 2007): 1–27.

Billing, Christian M. *Masculinity, Corporality and the English Stage, 1580–1635*. Farnham, UK: Ashgate, 2008.

Black, Ian. "NSA Spying Scandal: What We Have Learned." *The Guardian*, June 10, 2013. https://www.theguardian.com/world/2013/jun/10/nsa-spying-scandal-what-we-have-learned.

Blackstone, William. *The Oxford Edition of Blackstone's Commentaries on the Laws of England: Book II: Of the Rights of Things*. Oxford: Oxford University Press, 2016.

Blackwell, Bonnie. "The Pleasures of Geometry: Math and Melancholia in Rousseau's *La Nouvelle Helóïse*." *European Romantic Review* 15, no. 1 (March 2004): 89–112.

Blair, Ann. "Natural Philosophy and the 'New Science.'" In *The Cambridge History of Literary Criticism*, edited by Glyn P. Norton, 449–457. Cambridge: Cambridge University Press, 1999.

Blair, Ann, and Peter Stallybrass. "Mediating Information, 1450–1800." In *This Is Enlightenment*, edited by Clifford Siskin and William B. Warner, 139–163. Chicago: University of Chicago Press, 2010.

Blake, William. *Milton*. In *The Complete Illuminated Books*, 431–446. New York: Thames and Hudson, 2000.

Blewett, David. "The Double Time-Scheme of *Roxana*: Further Evidence." *Studies in Eighteenth-Century Culture* 13 (1984): 19–28.

Boehrer, Bruce. "Alsemero's Closet: Privacy and Interiority in *The Changeling*." *Journal of English and Germanic Philology* 96, no. 3 (July 1997): 349–368.

Boers, Elroy, Mohammad H. Afzali, Nicola Newton, and Patricia Conrod. "Association of Screen Time and Depression in Adolescence." *JAMA Pediatrics* 173, no. 9 (2019): 853–859.

Bonneuil, Christophe. "The Geological Turn: Narratives of the Anthropocene." In *The Anthropocene and the Global Environmental Crisis: Rethinking Modernity in a New Epoch*, edited by Clive Hamilton, Christophe Bonneuil, and François Gemenne, 17–31. Abingdon, UK: Routledge, 2015.

Borushko, Matthew. "Aesthetics of Nonviolence: Shelley, Adorno, Rancière." In *The Politics of Shelley: History, Theory, Form*, Romantic Circles PRAXIS, October 2015. https://romantic-circles.org/praxis/shelley_politics/praxis.2015.shelley_politics.borushko.html.

Bourne, John. *A Treatise on the Steam Engine in Its Application to Mines, Mills, Steam Navigation, and Railways*. London, 1846.

Boyle, John. *Remarks on the Life and Writings of Dr. Jonathan Swift*. London: A. Millar, 1752. Reprint, edited by Joao Froes, Newark: University of Delaware Press, 2000.

Brant, Clare. *Balloon Madness: Flights of the Imagination, 1783–1786*. Woodbridge, UK: Boydell Press, 2017.

Brooks, Peter. *Reading for the Plot: Design and Intention in Narrative*. New York: Knopf, 1984.

Bruenig, Matt. "Nickel-and-Dime Socialism." Medium, February 2017. https://medium.com/@MattBruenig/nickel-and-dime-socialism-47fcec406295.

Buci-Glucksmann, Christine. *The Madness of Vision: On Baroque Aesthetics*. Translated by Dorothy Z. Baker. Athens: Ohio University Press, 2013.

Bullard, Paddy, and James McLaverty. Introduction to *Jonathan Swift and the Eighteenth-Century Book*, edited by Paddy Bullard and James McLaverty, 1–28. Cambridge: Cambridge University Press, 2013.

Bullard, Rebecca. "Politics, History, and the *Robinson Crusoe* Story." In *Cambridge Companion to "Robinson Crusoe,"* edited by John Richetti, 84–96. Cambridge: Cambridge University Press, 2018.

Bunn, James. *Wave Forms: A Natural Syntax for Rhythmic Language*. Stanford, CA: Stanford University Press, 2002.

Burns, Russel. *Communications: An International History of the Formative Years*. London: Institute of Electrical Engineers, 2004.

Burson, Jeffrey D., and Anton M. Matytsin. Introduction to *The Skeptical Enlightenment: Doubt and Certainty in the Age of Reason*, edited by Jeffrey D. Burson and Anton M. Matytsin, 1–20. Liverpool: Liverpool University Press, 2019.

Butler, Judith. *Gender Trouble*. 10th anniversary ed. London: Taylor and Francis, 1999.

Butler, Marilyn. *Romantics, Rebels, and Reactionaries: English Literature and Its Background, 1760–1830*. Oxford: Oxford University Press, 1985.

Caldwell, Roy C. "*Tristram Shandy*, Bachelor Machine." *The Eighteenth Century* 34, no. 2 (Summer 1993): 103–114.

Calè, Luisa. "Extra-Illustration and Ephemera: Altered Books and the Alternative Forms of the Fugitive Page." *Eighteenth-Century Life* 44, no. 2 (2020): 111–135.

Campbell, Jill. "'I Am No Giant': Horace Walpole, Heterosexual Incest, and Love among Men." *The Eighteenth Century* 39, no. 3 (1998): 238–260.

Cardwell, D.S.L. "Science, Technology, and Industry." In *The Ferment of Knowledge: Studies in the Historiography of Eighteenth-Century Science*, edited by Roy Porter and G. S. Rousseau, 449–484. Cambridge: Cambridge University Press, 1980.

Carter, Ian. *Railways and Culture in Britain: The Epitome of Modernity.* Manchester: Manchester University Press, 2001.

Casid, Jill H. *Scenes of Projection: Recasting the Enlightenment Subject.* Minneapolis: University of Minnesota Press, 2014.

Castle, Terry. "Phantasmagoria: Spectral Technology and the Metaphorics of Modern Reverie." *Critical Inquiry* 15, no. 1 (1988): 26–61.

Cavendish, Margaret. *Observations upon Experimental Philosophy.* Edited by Eileen O'Neill. Cambridge: Cambridge University Press, 2001.

Chakravorty, Swapan. *Society and Politics in the Plays of Thomas Middleton.* Oxford: Clarendon Press, 1996.

Chamberlayne, Thomas. *The Compleat Midvvife's Practice Enlarged.* [London,] 1659.

Chandler, James. *England in 1819: The Politics of Literary Culture and the Case of Romantic Historicism.* Chicago: University of Chicago Press, 1998.

Chico, Tita. *The Experimental Imagination: Literary Knowledge and Science in the British Enlightenment.* Stanford, CA: Stanford University Press, 2018.

Cipolla, Carlo M. *Clocks and Culture, 1300–1700.* New York: Walker and Company, 1967.

Clayton, Jay. *Charles Dickens in Cyberspace: The Afterlife of the Nineteenth Century in Postmodern Culture.* Oxford: Oxford University Press, 2003.

Cody, Lisa Forman. "The Politics of Reproduction: From Midwives' Alternative Public Sphere to the Public Spectacle of Man-Midwifery." *Eighteenth-Century Studies* 32, no. 4 (Summer 1999): 477–495.

Cohen, Adam Max. *Shakespeare and Technology: Dramatizing Early Modern Technological Revolutions.* New York: Palgrave Macmillan, 2006.

———. *Technology and the Early Modern Self.* New York: Palgrave Macmillan, 2009.

Cohen, Ashley. "Wage Slavery, Oriental Despotism, and Global Labor: Management in Maria Edgeworth's *Popular Tales.*" *The Eighteenth Century* 55, nos. 2–3 (2014): 193–215.

Cohen, G. A. *Karl Marx's Theory of History: A Defence.* Princeton, NJ: Princeton University Press, 2000.

Congreve, William. *The Works of William Congreve.* Edited by D. F. McKenzie. 3 vols. Oxford: Oxford University Press, 2011.

Conrad, Peter. *Shandyism: The Character of Romantic Irony.* Oxford: Basil Blackwell, 1978.

Cook, Harold J. *Matters of Exchange: Commerce, Medicine, and Science in the Dutch Golden Age.* New Haven, CT: Yale University Press, 2007.

Cookson, Gillian. *The Age of Machinery: Engineering the Industrial Revolution, 1770–1850.* Suffolk, UK: Boydell, 2018.

Coppola, Al. *The Theater of Experiment: Staging Natural Philosophy in Eighteenth-Century Britain.* Oxford: Oxford University Press, 2016.

Courier and Evening Gazette (London). November 21, 1784. Issue 730.

———. November 26, 1794. Issue 734.

Cowper, William. *The Anatomy of Humane Bodies.* London, 1698.

Coykendall, Abby. "Chance Enlightenments, Choice Superstitions: Walpole's Historic Doubts and Enlightenment Historicism." *The Eighteenth Century* 54, no. 1 (2013): 53–70.

Crane, R. S. "The Houyhnhnms, the Yahoos, and the History of Ideas." In *Reason and the Imagination: Studies in the History of Ideas, 1600–1800,* edited by J. A. Mazzeo, 231–253. New York: Columbia University Press, 1962.

Critical Review, January 1765. In *Horace Walpole: The Critical Heritage,* edited by Peter Sabor, 68–69. New York: Routledge, 1995.

Crooke, Helkiah. *Mikrokosmographia.* London, 1615.

Crutzen, Paul, and Eugene Stoermer. "The 'Anthropocene.'" *IGBP Newsletter* 41 (2000): 17–18.

Culpeper, Nicholas. *A Directory for Midwives*. London, 1651.
———. *A Directory for Midwives*. Rev. ed. London, 1716.
Curran, Stuart. *Shelley's Annus Mirabilis: The Maturing of an Epic Vision*. San Marino, CA: Huntington Library Press, 1975.
Curtis, Adam, dir. *Hypernormalisation*. London: BBC, BBC iPlayer, 2016.
Dackerman, Susan. Introduction to *Prints and the Pursuit of Knowledge in Early Modern Europe*, edited by Susan Dackerman, 19–35. Cambridge, MA: Harvard Art Museums, 2018.
Dacome, Lucia. *Malleable Anatomies: Models, Makers, and Material Culture in Eighteenth-Century Italy*. Oxford: Oxford University Press, 2017.
———. "Women, Wax and Anatomy in the 'Century of Things.'" *Renaissance Studies* 21, no. 4 (2007): 522–550.
Daly, Nicholas. *Literature, Technology, and Modernity, 1860–2000*. Cambridge: Cambridge University Press, 2004.
Davies, Jeremy. *The Birth of the Anthropocene*. Berkeley: University of California Press, 2016.
Day, Robert Adams. *Told in Letters: Epistolary Fiction before Richardson*. Ann Arbor: University of Michigan Press, 1966.
"Defence of Coach-Building, and a Few Words on Steam-Carriages." *Mechanics Magazine* 18 (1833): 209–210.
Defoe, Daniel. *The Fortunate Mistress; or, a History of the Life and Vast Variety of Fortunes of Mademoiselle de Beleau . . . Being the Person Known by the Name of the Lady Roxana, in the Time of King Charles II*. 2 vols. Oxford: Basil Blackwell / Shakespeare Head Press, 1927.
———. *A Journal of the Plague Year*. Edited by Paula R. Backscheider. New York: Norton, 1992.
———. *The Life & Strange Surprising Adventures of Robinson Crusoe, of York*. 3 vols. Oxford: Basil Blackwell / Shakespeare Head Press, 1927.
———. *A Review of the Affairs of France*. Edited by John McVeagh. 9 vols. London: Pickering and Chatto, 2003.
———. *A Tour through the Whole Island of Great Britain*. Edited by Pat Rodgers. London: Penguin, 2005.
Deleuze, Gilles. *The Fold: Leibniz and the Baroque*. Translated by Tom Conley. Minneapolis: University of Minnesota Press, 1993.
De Man, Paul. *Blindness and Insight: Essays in the Rhetoric of Contemporary Criticism*. London: Routledge, 2005.
———. "Shelley Disfigured." In *Deconstruction and Criticism*, edited by Harold Bloom, 32–61. New York: Continuum, 2004.
Dennis, John. *The Plays of John Dennis*. Edited by J. W. Johnson. New York: Garland Publishing, 1980.
Desaguliers, John Theophilus. *A Course of Experimental Philosophy*. Vol. 1. London: W. Innys, 1745.
Descartes, René. *Meditations on First Philosophy with Selections from the Objections and Replies*. Translated by Michael Moriarty. New York: Oxford University Press, 2008.
Dibdin, Charles. *A Collection of Songs, Selected from the Works of Mr. Dibdin. To Which Are Added, the Newest and Most Favourite American Patriotic Songs*. Philadelphia: Printed by J. Bioren for H. & P. Price, and sold by J. Rice, Baltimore, 1799.
Dick, Alex. *Romanticism and the Gold Standard: Money, Literature, and Economic Debate in Britain, 1790–1830*. London: Palgrave Macmillan, 2013.
di Marco, Corolini. *The Flying Island, &c. Being a Key to Gulliver's Voyage to Laputa, Balnibarbi, Glubbdubdrib, Luggnag, and Japan*. London: Edmund Curll, 1726.

Dinshaw, Carolyn. *How Soon Is Now? Medieval Texts, Amateur Readers, and the Queerness of Time*. Durham, NC: Duke University Press, 2012.

Dolan, Jill. *The Feminist Spectator as Critic*. Ann Arbor, MI: UMI Research Press, 1988.

Douglas, Aileen. *Work in Hand: Script, Print, and Writing, 1690–1840*. Oxford: Oxford University Press, 2017.

Dowd, Michelle M. "Delinquent Pedigrees: Revision, Lineage, and Spatial Rhetoric in *The Duchess of Malfi*." *English Literary Renaissance* 39, no. 3 (2009): 499–526.

Doyle, Laura. *Freedom's Empire: Race and the Rise of the Novel in Atlantic Modernity, 1640–1940*. Durham, NC: Duke University Press, 2008.

Drury, Joseph. "Literature and Science in Enlightenment Britain: New Directions." *Literature Compass* 14, no. 6 (2017). https://compass.onlinelibrary.wiley.com/doi/10.1111/lic3.12395.

———. *Novel Machines: Technology and Narrative Form in Enlightenment Britain*. Oxford: Oxford University Press, 2017.

Duffy, Enda. *The Speed Handbook: Velocity, Pleasure, Modernism*. Durham, NC: Duke University Press, 2009.

Eavis, Anna, and Michael Peover. *Horace Walpole's Painted Glass at Strawberry Hill*. London: British Society of Master Glass Painters, 1995.

Edelman, Lee. *No Future: Queer Theory and the Death Drive*. Durham, NC: Duke University Press, 2004.

———. "The Pathology of the Future, or the Endless Triumphs of Life." In *Constellations of a Contemporary Romanticism*, edited by Jacques Khalip and Forest Pyle, 34–42. New York: Fordham University Press, 2016.

Edgerton, David. "From Innovation to Use: Ten Eclectic Theses on the Historiography of Technology." *History and Technology* 16, no. 2 (1999): 111–136.

———. *The Shock of the Old: Technology and Global History since 1900*. London: Penguin, 2019.

Edgeworth, Maria. *Popular Tales*. London: Printed for J. Johnson, St. Paul's Church-yard, by C. Mercier and Co., 1804.

Edgeworth, Maria, and Richard Lovell Edgeworth. *Memoirs of Richard Lovell Edgeworth, Esq*. London: Printed for R. Hunter, 1820.

Edgeworth, Richard Lovell. "An Essay on the Art of Conveying Secret and Swift Intelligence." *Transactions of the Royal Irish Academy* 6 (1797): 95–139.

———. *A Letter to the Right Hon. The Earl of Charlemont, on the Tellograph, and on the Defence of Ireland*. London: Printed for J. Johnson, in St. Paul's Church Yard, 1797.

Empson, William. *Seven Types of Ambiguity*. New York: New Directions, 1966.

Erwin, Timothy. "William Hogarth and the Aesthetics of Nationalism." *Huntington Library Quarterly* 64, no. 3 (2001): 383–411.

Etmullerus, Michael. *Etmullerus Abridg'd*. London, 1703.

An Explanation of the Nature of Equation of Time, and Use of the Equation Table for Adjusting Clocks and Watches to the Motion of the Sun. London: F. Clay, 1731.

Faflak, Joel. "Dancing in the Dark with Shelley." In *Constellations of a Contemporary Romanticism*, edited by Jacques Khalip and Forest Pyle, 147–161. New York: Fordham University Press, 2016.

———. "Right to Romanticism." *European Romantic Review* 27, no. 3 (2016): 285–300.

Fairclough, Mary. "The *Telegraph*: Radical Transmission in the 1790s." *Eighteenth-Century Life* 37, no. 2 (2013): 26–52.

Farquhar, George. *The Works of George Farquhar*. Edited by Shirley Strum Kenny. 2 vols. Oxford: Clarendon Press, 1988.

Farr, Jason. *Novel Bodies: Disability and Sexuality in Eighteenth-Century Literature*. Lewisburg, PA: Bucknell University Press, 2019.

Feenberg, Andrew. *Questioning Technology*. London: Routledge, 1999.

Felski, Rita. *The Limits of Critique*. Chicago: University of Chicago Press, 2015.

———. *The Uses of Literature*. Oxford: Blackwell, 2008.

Ferguson, Adam. *An Essay on the History of Civil Society, 1767*. Piscataway, NJ: Transaction Publishers, 1980.

Fissell, Mary E. "Remaking the Maternal Body in England, 1680–1730." *Journal of the History of Sexuality* 26, no. 1 (January 2017): 114–139.

———. *Vernacular Bodies: The Politics of Reproduction in Early Modern England*. Oxford: Oxford University Press, 2004.

Floyd-Wilson, Mary. *Occult Knowledge, Science, and Gender on the Shakespearean Stage*. Cambridge: Cambridge University Press, 2013.

Foot, Paul. *Red Shelley*. London: Sidgwick and Jackson, 1980.

Forman, Paul. "The Primacy of Science in Modernity, of Technology in Postmodernity, and of Ideology in the History of Technology." *History and Technology* 23, nos. 1–2 (2007): 1–152.

Foucault, Michel. *The Birth of Biopolitics: Lectures at the Collège de France, 1978–79*. Translated by Graham Burchell. London: Palgrave Macmillan, 2008.

———. *The Order of the Things: An Archeology of the Human Sciences*. Anonymous translation. New York: Vintage Books, 1970.

———. *Power: Essential Works of Foucault, 1954–84*. Translated Robert Hurley. New York: New Press, 2000.

Fox, Christopher. "Introduction: How to Prepare a Noble Savage: The Spectacle of Human Science." In *Inventing Human Science: Eighteenth-Century Domains*, edited by Roy Porter Fox and Robert Wokler, 1–30. Berkeley: University of California Press, 1995.

Fraser, Nancy. *Fortunes of Feminism: From State-Managed Capitalism to Neoliberal Crisis*. London: Verso, 2013.

Freeman, Elizabeth. *Time Binds: Queer Temporalities, Queer Histories*. Durham, NC: Duke University Press, 2010.

Freeman, Michael J. *Railways and the Victorian Imagination*. New Haven, CT: Yale University Press, 1999.

Fuller, John. Introduction to *John Gay: Dramatic Works*, edited by John Fuller, 1–76. Oxford: Clarendon Press, 1983.

Fyfe, Paul. "On the Opening of the Liverpool and Manchester Railway, 1830." *BRANCH: Britain, Representation and Nineteenth-Century History*, December 2012. http://www.branchcollective.org/?ps_articles=paul-fyfe-on-the-opening-of-the-liverpool-and-manchester-railway-1830.

Gaedtke, Andrew. *Modernism and the Machinery of Madness: Psychosis, Technology, and Narrative Worlds*. Cambridge: Cambridge University Press, 2017.

Gal, Ofer, and Raz Chen-Morris. *Baroque Science*. Chicago: University of Chicago Press, 2013.

Gang, Joshua. "Behaviorism and the Beginnings of Close Reading." *ELH* 78, no. 1 (2011): 1–25.

García Santo-Tomás, Enrique. *La musa refractada: Literatura y óptica en la España del Barroco*. Madrid: Iberoamericana, 2014.

Gay, John. *Three Hours after Marriage*. 1717. In *John Gay: Dramatic Works*, edited by John Fuller, vol. 1, 207–263. Oxford: Clarendon Press, 1983.

The Gentleman's Magazine 64 (November 1794).

Ghosh, Amitav. *The Great Derangement: Climate Change and the Unthinkable*. Chicago: University of Chicago Press, 2016.

Gibson, Thomas. *The Anatomy of Humane Bodies Epitomized*. London, 1682.

Giedion, Siegfried. *Mechanization Takes Command: A Contribution to Anonymous History*. Minneapolis: University of Minnesota Press, 2013.

Gilmartin, Kevin. *Writing against Revolution: Literary Conservatism in Britain, 1790–1832*. Cambridge: Cambridge University Press, 2006.

Girten, Kristin M. "Mingling with Matter: Tactile Microscopy and the Philosophic Mind in Brobdingnag and Beyond." *The Eighteenth Century* 54, no. 4 (2013): 497–520.

Glennie, Paul, and Nigel Thrift. *Shaping the Day: A History of Timekeeping in England and Wales, 1300–1800*. Oxford: Oxford University Press, 2009.

Glissant, Édouard. *The Poetics of Relation*. Translated by Betsy Wing. Ann Arbor: University of Michigan Press, 1997.

Goldstein, Amanda Jo. *Sweet Science: Romantic Materialism and the New Logics of Life*. Chicago: University of Chicago Press, 2017.

Golway, Terry. *Machine Made: Tammany Hall and the Creation of Modern American Politics*. New York: Norton, 2014.

Graham, Harvey. *Eternal Eve*. London: William Heinemann Medical Books, 1950.

Graves, R. B. "'The Duchess of Malfi' at the Globe and Blackfriars." *Renaissance Drama*, n.s., no. 9 (1978): 193–209.

Great Britain. *An Act for Providing a Public Reward for Such Person or Persons as Shall Discover the Longitude at Sea*. London: John Baskett, assigns of Thomas Newcomb, and Henry Hills, 1714.

Griffiths, Trevor, Philip A. Hunt, and Patrick K. O'Brien. "Inventive Activity in the British Textile Industry, 1700–1800." *Journal of Economic History* 52, no. 4 (1992): 881–906.

Gubar, Susan. "The Female Monster in Augustan Satire." *Signs* 3, no. 2 (Winter 1977): 380–394.

Guerrini, Anita. *The Courtiers' Anatomists: Animals and Humans in Louis XIV's Paris*. Chicago: University of Chicago Press, 2015.

Gwilliam, Tassie. "Female Fraud: Counterfeit Maidenheads in the Eighteenth Century." *Journal of the History of Sexuality* 6, no. 4 (April 1996): 518–548.

Habermas, Jürgen. *The Philosophical Discourse of Modernity*. Cambridge, MA: MIT Press, 1990.

Haggerty, George. *Horace Walpole's Letters: Masculinity and Friendship in the Eighteenth Century*. Lewisburg, PA: Bucknell University Press, 2011.

Hahn, Barbara. *Technology in the Industrial Revolution*. Cambridge: Cambridge University Press, 2020.

Hahn, Roger. "The Application of Science to Society: The Societies of the Arts." *Studies on Voltaire and the Eighteenth Century* 25 (1963): 829–836.

Halberstam, Jack. *In a Queer Time and Place: Transgender Bodies, Subcultural Lives*. New York: New York University Press, 2005.

Hankins, Thomas L., and Robert J. Silverman. *Instruments and the Imagination*. Princeton, NJ: Princeton University Press, 1995.

Haraway, Donna. "Manifesto for Cyborgs: Science, Technology, and Socialist Feminism in the Late Twentieth Century." In *Simians, Cyborgs, and Women: The Reinvention of Nature*, 149–181. New York: Routledge, 1991.

Hardt, Michael, and Antonio Negri. *Commonwealth*. Cambridge, MA: Harvard University Press, 2009.

Harkness, Deborah. "'Strange' Ideas and 'English' Knowledge: Natural Science Exchange in Elizabethan London." In *Merchants & Marvels: Commerce, Science, and Art in Early Modern Europe*, edited by Pamela H. Smith and Paula Findlen, 137–162. New York: Routledge, 2002.

Harlizius-Kluck, Ellen. "Weaving as Binary Art and the Algebra of Patterns." *TEXTILE* 15, no. 2 (2017): 176–197.

Harol, Corinne. *Enlightened Virginity in Eighteenth-Century Literature*. New York: Macmillan, 2006.

Harris, Anne F. "Glazing and Glossing: Stained Glass as Literary Interpretation." *Journal of Glass Studies* 56 (2014): 303–316.

Harris, Jocelyn. *Satire, Celebrity, and Politics in Jane Austen*. Lewisburg, PA: Bucknell University Press, 2017.

Harvey, Karen. "The Century of Sex? Gender, Bodies, and Sexuality in the Long Eighteenth Century." *The Historical Journal* 45, no. 4 (December 2002): 899–916.

Hawley, Judith. "*Tristram Shandy*, Learned Wit, and Enlightenment Knowledge." In *The Cambridge Companion to Laurence Sterne*, edited by Thomas Keymer, 34–48. Cambridge: Cambridge University Press, 2009.

Hayles, N. Katherine. *How We Became Posthuman: Virtual Bodies in Cybernetics, Literature, and Informatics*. Chicago: University of Chicago Press, 1999.

Hazlitt, William. "Review of *Biographia Literaria*." *Edinburgh Review* 28 (1817): 488–515.

———. "Review of Shelley's *Posthumous Poems*." In *Shelley: The Critical Heritage*, edited by James E. Barcus, 335–345. London: Routledge and Kegan Paul, 1975.

[Hearne, Mary]. *The Female Deserters: A Novel*. London: J. Roberts, 1718–1719.

———. *The Lover's Week; or, The Six Days Adventures of Philander and Amaryllis*. London: E. Curll and R. Francklin, 1718.

Hegel, G.W.F. *The Early Political Writings of the German Romantics*. Edited and translated by Frederick C. Baiser. Cambridge: Cambridge University Press, 1999.

———. *Hegel's Philosophy of Right*. Translated by T. M. Knox. Oxford: Clarendon Press, 1978.

Heidegger, Martin. "The Question Concerning Technology." In *The Question Concerning Technology and Other Essays*, edited and translated by William Lovitt, 3–35. New York: Garland Publishing, 1977.

Heise, Ursula K. "Lost Dogs, Last Birds, and Listed Species: Cultures of Extinction." *Configurations* 18, nos. 1–2 (Winter 2010): 49–72.

Henderson, Andrea. "Death on the Stage, Death of the Stage: The Antitheatricality of *The Duchess of Malfi*." *Theatre Journal* 42, no. 2 (1990): 194–207.

Hershinow, Stephanie Insley. "The Incest Plot: Marriage, Closure, and the Novel's Endogamy." *The Eighteenth Century* 61, no. 2 (2020): 149–164.

Hilaire-Pérez, Liliane, and Catherine Verna. "Dissemination of Technical Knowledge in the Middle Ages and the Early Modern Era." *Technology and Culture* 47 (July 2006): 536–565.

Hobbes, Thomas. *Leviathan*. Edited by Richard Tuck. Cambridge: Cambridge University Press, 1996.

———. *Leviathan*. Rev. ed. Toronto: Broadview Press, 2011.

Hogarth, William. *The Analysis of Beauty*. New Haven, CT: Yale University Press, 1997.

Hogle, Jerrold. *Shelley's Process: Radical Transference and the Development of His Major Works*. New York: Oxford University Press, 1989.

Home, Henry, Lord Kames. *The Elements of Criticism*. 6th ed. Edited by Peter Jones. 2 vols. Indianapolis: Liberty Fund, 2005.

Hone, William. *The Political House That Jack Built*. Romantic Circles: Editions, 1998. https://romantic-circles.org/editions/hone/jacktext.htm.

———. *The Right Divine of Kings to Govern Wrong!* London: William Hone, 1821.

Hooke, Robert. *Micrographia; or, Some Physiological Descriptions of Minute Bodies Made by Magnifying Glasses*. London: Martyn and Allestree, 1665.

Howse, Derek. *Greenwich Time and the Discovery of the Longitude*. Oxford: Oxford University Press, 1980.

Hughes, John. *The Spectator*, no. 220 (November 12, 1711). Edited by Richard Steele. In *The Spectator*, vol. 3, edited by George A. Aitken, 246–248. New York: Longmans, Green, and Co., 1898.

Hunt, Leigh. "Disturbances at Manchester." *The Examiner*, August 22, 1819.

Hutcheson, Francis. *An Inquiry into the Original of Our Ideas of Beauty and Virtue*. London, 1726.

Hyman, Wendy Beth. "'Deductions from Metaphors': Figurative Truth, Poetical Language, and Early Modern Science." In *The Palgrave Handbook of Early Modern Literature and Science*, edited by Howard Marchitello and Evelyn Tribble, 27–48. London: Palgrave Macmillan, 2017.

Ingold, Tim. *Lines: A Brief History*. London: Routledge, 2007.

Jackson, Zakiyyah Iman. *Becoming Human: Matter and Meaning in an Antiblack World*. New York: New York University Press, 2020.

Jameson, Fredric. *Postmodernism, or, The Cultural Logic of Late Capitalism*. Durham, NC: Duke University Press, 1991.

———. *Representing Capital: A Commentary on Volume One*. London: Verso, 2011.

———. *A Singular Modernity: Essay on the Ontology of the Present*. London: Verso, 2013.

Jardine, Lisa, and Michael Silverthorne. Introduction to *The New Organon*, by Francis Bacon, vii–xxviii. Cambridge: Cambridge University Press, 2000.

Jenkins, David T. *The Cambridge History of Western Textiles*. Cambridge: Cambridge University Press, 2003.

John S. McCain National Defense Authorization Act for Fiscal Year 2019. Public Law 115-232. 115th Congress. Washington, DC: U.S. Government Publishing Office, August 13, 2018.

Johnson, Samuel. *A Dictionary of the English Language: In Which the Words Are Deduced from Their Originals, Explained in Their Different Meanings, and Authorized by the Names of Writers in Whose Works They are Found*. Edited by Alexander Chalmers. London: C. and J. Rivington, 1824.

Jones, David J. *Gothic Machine: Textualities, Pre-cinematic Media and Film in Popular Visual Culture, 1670–1910*. Cardiff: University of Wales Press, 2011.

Jones, Matthew L. *Reckoning with Matter: Calculating Machines, Innovation, and Thinking about Thinking from Pascal to Babbage*. Chicago: University of Chicago Press, 2016.

Kantor, Jamison. "Horace Walpole and the Fate of Finance." *The Eighteenth Century* 58, no. 2 (2017): 135–155.

Kassell, Lauren. "Medical Understandings of the Body, *c.* 1500–1750." In *The Routledge History of Sex and the Body*, edited by Sarah Toulalan and Kate Fisher, 57–74. London: Routledge, 2013.

Kaufman, Robert. "Legislators of the Post-Everything World: Shelley's 'Defence' of Adorno." *ELH* 63, no. 3 (1996): 707–733.

———. "The Work of Romanticism in the Age of Mechanical Postmodernism." *European Romantic Review* 12, no. 2 (2001): 237–245.

Kelly, Kathleen Coyne. *Performing Virginity and Testing Chastity in the Middle Ages*. London: Routledge, 2000.

Kennedy, Robert C. "Probe Away." *HarpWeek*. Accessed June 2019. https://www.harpweek.com/09Cartoon/BrowseByDateCartoon.asp?Month=March&Date=18.

Ketabgian, Tamara S. *The Lives of Machines: The Industrial Imaginary in Victorian Literature and Culture*. Ann Arbor: University of Michigan Press, 2016.

Kim, Dorothy. "Introduction to Literature Compass Special Cluster: Critical Race and the Middle Ages." *Literature Compass* 16, nos. 9–10 (2019): 1–16.

King, Kathryn R. "The Novel before Novels (with a Glance at Mary Hearne's Fables of Desertion)." In *Eighteenth-Century Genre and Culture: Serious Reflections on Occasional Forms, Essays in Honor of J. Paul Hunter*, edited by Dennis Todd and Cynthia Wall, 36–57. Newark: University of Delaware Press, 2001.

Kirchgaessner, Stephanie, Paul Lewis, David Pegg, Sam Cutler, Nina Lakhani, and Michael Safi. "Revealed: Leak Uncovers Global Abuse of Cyber-Surveillance Weapon." *The Guardian*, July 18, 2021. https://www.theguardian.com/world/2021/jul/18/revealed-leak-uncovers-global-abuse-of-cyber-surveillance-weapon-nso-group-pegasus.

Krajewski, Markus. *Paper Machines: About Cards & Catalogs, 1548–1929*. Translated by Peter Krapp. Cambridge, MA: MIT Press, 2011.

Kramnick, Jonathan. "Against Literary Darwinism." *Critical Inquiry* 37, no. 2 (2011): 315–347.

Kuiken, Kir. *Imagined Sovereignties: Toward a New Political Romanticism*. New York: Fordham University Press, 2014.

Labour Party. *For the Many, Not the Few: The Labour Party Manifesto 2017*. October 2017. https://labour.org.uk/wp-content/uploads/2017/10/labour-manifesto-2017.pdf.

Lähteenmäki, Vili. "Locke and Active Perception." In *Active Perception in the History of Philosophy: From Plato to Modern Philosophy*, edited by José Filipe Silva and Mikko Yrjönsuuri, 223–240. New York: Springer, 2014.

Landes, David. *Revolution in Time: Clocks and the Making of the Modern World*. Rev. ed. Cambridge, MA: Belknap Press of Harvard University Press, 2000.

Landy, Joshua. *How to Do Things with Fictions*. Stanford, CA: Stanford University Press, 2012.

Laqueur, Thomas. *Making Sex: Body and Gender from the Greeks to Freud*. Cambridge, MA: Harvard University Press, 1990.

Latour, Bruno. *Reassembling the Social: An Introduction to Actor-Network Theory*. Oxford: Oxford University Press, 2005.

———. *We Have Never Been Modern*. Translated by Catherine Porter. Cambridge, MA: Harvard University Press, 1993.

Leake, John. *Introduction to the Theory and Practice of Midwifery*. London, 1787.

LeMenager, Stephanie. "Climate Change and the Struggle for Genre." In *Anthropocene Reading: Literary History in Geologic Times*, edited by Tobias Menely and Jesse Oak Taylor, 220–238. University Park: Pennsylvania State University Press, 2017.

Letellier, Robert Ignatius. *The English Novel, 1700–1740: An Annotated Bibliography*. Westport, CT: Greenwood Press, 2002.

Letter reproduced in *Three Hours after Marriage*. Dublin, 1761.

Levine, Caroline. *Forms: Whole, Rhythm, Hierarchy, Network*. Princeton, NJ: Princeton University Press, 2015.

Lewis, Michael. *The Big Short: Inside the Doomsday Machine*. New York: Norton, 2011.

Lewis, Simon L., and Mark A. Maslin. *The Human Planet: How We Created the Anthropocene*. New Haven CT: Yale University Press, 2018.

Lezra, Jacques. *Unspeakable Subjects: The Genealogy of the Event in Early Modern Europe*. Stanford, CA: Stanford University Press, 1997.

Lieberman, Jennifer L. "Finding a Place for Technology." *JLS: Journal of Literature and Science* 10, no. 1 (2017): 26–31.

Locke, John. *An Essay Concerning Human Understanding*. Edited by Peter H. Nidditch. Oxford: Clarendon Press, 1975.

Loewenstein, Joseph. *The Author's Due: Printing and the Prehistory of Copyright*. Chicago: University of Chicago Press, 2002.

Loretelli, Rosamaria. "The Aesthetics of Empiricism and the Origin of the Novel." *The Eighteenth Century* 41, no. 2 (Summer 2000): 83–109.

Loughlin, Marie H. *Hymeneutics: Interpreting Virginity on the Early Modern Stage.* Lewisburg, PA: Bucknell University Press, 1997.

Löwy, Michael. *Ecosocialism: A Radical Alternative to Capitalist Catastrophe.* Chicago: Haymarket Books, 2015.

Lupton, Christina. "Contingency, Codex, the Eighteenth-Century Novel." *ELH* 81, no. 4 (Winter 2014): 1173–1192.

Luttfring, Sara D. "Bodily Narratives and the Politics of Virginity in *The Changeling* and the Essex Divorce." *Renaissance Drama* 39, no. 1 (2011): 97–128.

Lynall, Gregory. *Swift and Science: The Satire, Politics, and Theology of Natural Knowledge, 1690–1730.* New York: Springer, 2012.

Macaulay, Thomas Babington. *The History of England from the Accession of James II.* Vol. 2. London: Longman, Brown, Green, and Longmans, 1849.

Mack, Ruth. "Hogarth's Practical Aesthetics." In *Mind, Body, Motion, Matter: Eighteenth-Century British and French Literary Perspectives,* edited by Mary Helen McMurray and Alison Conway, 21–46. Toronto: University of Toronto Press, 2016.

———. "Paper Castle, Paper Collection: Walpole's Extra-Illustrated Copy of the *Description of . . . Strawberry Hill.*" In *Horace Walpole's Strawberry Hill,* edited by Michael Snodin and Cynthia Roman, 107–115. New Haven, CT: Yale University Press, 2009.

MacKenzie, Scott. "My Chapter upon Lines: Motion, Deviation, and Lineation in Eighteenth-Century British Aesthetics." *Criticism* 61, no. 1 (2019): 1–26.

Majendie, Lewis. *An Account of Hedingham Castle, in the County of Essex.* London: John Nichols, 1796.

Malm, Andreas. *Fossil Capital: The Rise of Steam Power and the Roots of Global Warming.* London: Verso, 2016.

Mann, Jenny C. *The Trials of Orpheus: Poetry, Science, and the Early Modern Sublime.* Princeton, NJ: Princeton University Press, 2021.

Mann, Jenny C., and Debapriya Sarkar, eds. "Imagining Early Modern Scientific Forms." Special issue, *Philological Quarterly* 98, no. 1–2 (Winter/Spring 2019).

Mannheimer, Katherine. "The Scriblerian Stage and Page: *Three Hours after Marriage,* Pope's 'Minor' Poems, and the Problem of Genre-History." *Comparative Drama* 43, no. 1 (Spring 2009): 63–88.

Maravall, José Antonio. *La cultural del barroco: Análisis de una estructura histórica.* Barcelona: Editorial Planeta, 1975.

Marchitello, Howard, and Evelyn Tribble, eds. *The Palgrave Handbook of Early Modern Literature and Science.* London: Palgrave Macmillan, 2017.

Margócsy, Dániel. *Commercial Visions: Science, Trade, and Visual Culture in the Dutch Golden Age.* Chicago: University of Chicago Press, 2014.

Marks, Richard. "The Reception and Display of Northern European Roundels in England." *Gesta* 37, no. 2 (1998): 217–224.

Marsden, Ben. *Watt's Perfect Engine: Steam and the Age of Invention.* New York: Columbia University Press, 2002.

Marten, John. *Gonosologium Novum.* London, 1709.

Marx, Karl. *Capital.* Vol. 1. Translated by Ben Fowkes. London: Penguin, 1990.

Marx, Karl, and Frederick Engel. *Collected Works.* Vol. 11. New York: International Publishers, 1979.

Massey, Lyle. "Pregnancy and Pathology: Picturing Childbirth in Eighteenth-Century Obstetric Atlases." *The Art Bulletin* 87, no. 1 (March 2005): 73–91.

Maxwell, Lynn. *Wax Impressions, Figures, and Forms in Early Modern Literature: Wax Works.* New York: Springer, 2019.

———. “Wax Magic and *The Duchess of Malfi*.” *Journal for Early Modern Cultural Studies* 14, no. 3 (2014): 31–54.

Mazzio, Carla. “The Invisible Element in Art: Dürer, Shakespeare, Donne.” In *Vision and Its Instruments: Art, Science, and Technology in Early Modern Europe*, edited by Alina Alexandra Payne, 138–166. University Park: Pennsylvania State University Press, 2015.

McClean, Bethany, and Joseph Nocera. *All The Devils Are Here: The Hidden History of the Financial Crisis*. New York: Portfolio/Penguin, 2011.

McClellan, James E., III. *Science Reorganized: Scientific Societies in the Eighteenth Century*. New York: Columbia University Press, 1985.

McCorduck, Pamela. *Machines Who Think: A Personal Inquiry into the History and Prospects of Artificial Intelligence*. Natick, MA: A. K. Peters, 1979.

McGrath, Alice Tweedy. “Unaccountable Form: Queer Failure and Jane Barker’s Patchwork Method.” *The Eighteenth Century* 60, no. 4 (2019): 353–373.

McKenny, John. “Swift’s Prescience: A Polite Precursor of Corpus Linguistics.” *Journal of Language and Literature* 2, no. 1 (2003): 47–65.

McNeill, J. R., and Peter Engelke. *The Great Acceleration: An Environmental History of the Anthropocene since 1945*. Cambridge MA: Harvard University Press, 2016.

Meisel, Martin. *Chaos Imagined: Literature, Art, Science*. New York: Columbia University Press, 2016.

Mikhail, Alan. “Enlightenment Anthropocene.” *Eighteenth-Century Studies* 49, no. 2 (2016): 211–231.

Miller, David Philip. *James Watt, Chemist: Understanding the Origins of the Steam Age*. London: Pickering and Chatto, 2009.

Milton, John. *Complete Prose Works of John Milton*. Edited by Merritt Y. Hughes. Vol. 3. New Haven, CT: Yale University Press, 1962.

Misson, Henri. *Memoires et observations faites par un voyageur en Angleterre*. The Hague: Henri van Bulderen, 1698.

Mitchell, Robert. *Experimental Life: Vitalism in Romantic Science and Literature*. Baltimore: Johns Hopkins University Press, 2013.

Mitchell, W.J.T. *Articulate Images: The Sister Arts from Hogarth to Tennyson*. Minneapolis: University of Minnesota Press, 1983.

Moore, Jason. *Capitalism in the Web of Life: Ecology and the Accumulation of Capital*. London: Verso, 2015.

Morning Chronicle (London). November 25, 1794.

Morning Post (London). November 29, 1794.

Morning Post and Fashionable World (London). September 19, 1794.

Morton, Timothy. *Hyperobjects. Philosophy and Ecology after the End of the World*. Minneapolis: University of Minnesota Press, 2013.

Motte, Andrew. *A Treatise of the Mechanical Powers, Wherein the Laws of Motions and the Properties of Those Powers Are Explain’d and Demonstrated in an Easy and Familiar Method*. London: Motte, 1733.

Muirhead, James Patrick. *The Origin and Progress of the Mechanical Inventions of James Watt*. 3 vols. London: John Murray, 1854.

Müller, Simone M., and Heidi J. S. Tworek. “Imagined Use as a Category of Analysis: New Approaches to the History of Technology.” *History and Technology* 32, no. 2 (2016): 105–119.

Müller-Sievers, Helmut. *The Cylinder: Kinematics of the Nineteenth Century*. Berkeley: University of California Press, 2012.

Muñoz, José Esteban. *Cruising Utopia: The Then and There of Queer Futurity*. New York: New York University Press, 2009.

Muri, Allison. *The Enlightenment Cyborg: A History of Communications and Control in the Human Machine, 1660–1830*. Toronto: University of Toronto Press, 2007.

Nägele, Rainer. *Theater, Theory, Speculation: Walter Benjamin and the Scenes of Modernity*. Baltimore: Johns Hopkins University Press, 1991.

Newton, Isaac. "Author's Preface to the Reader." In *The Principia: Mathematical Principles of Natural Philosophy*, translated by I. Bernard Cohen and Anne Whitman, assisted by Julia Budenz (Berkeley: University of California Press, 2016), 382.

Nicholson, Michael. "Fugitive Pieces: Walpole, Byron, and Queer Time." *The Eighteenth Century* 60, no. 2 (2019): 139–162.

Nicolson, Marjorie. "The Microscope and English Imagination." *Smith College Studies in Modern Languages* 16, no. 4 (1935): 1–92.

Nicolson, Marjorie, and Nora M. Mohler. "The Scientific Background of Swift's Voyage to Laputa." *Annals of Science* 2, no. 3 (1937): 299–334.

Nicolson, Marjorie Hope, and David Stuart Rodes. Introduction to *The Virtuoso*, by Thomas Shadwell, xi–xxvi. Lincoln: University of Nebraska Press, 1966.

Norman, Don. *The Psychology of Everyday Things*. New York: Basic Books, 1988.

Nunn, Hillary M. *Staging Anatomies: Dissection and Spectacle in Early Stuart Tragedy*. Burlington, VT: Ashgate, 2005.

OED Online. Oxford: Oxford University Press. https://www.oed.com.

Offray de La Mettrie, Julien. *La Mettrie: Machine Man and Other Writings*. Edited by Ann Thomson. Cambridge: Cambridge University Press, 1996.

Ogée, Frédéric. "From Text to Image: William Hogarth and the Emergence of a Visual Culture in Eighteenth-Century England." In *Hogarth: Representing Nature's Machines*, edited by David Bindman, Frédéric Ogée, and Peter Wagner, 3–22. Manchester: Manchester University Press, 2001.

Ortiz, Ivan. "Romanticism and the Aesthetics of Modern Transport." PhD diss., Princeton University, 2013.

Otis, Laura. *Networking: Communicating with Bodies and Machines in the Nineteenth Century*. Ann Arbor: University of Michigan Press, 2011.

Owens, Margaret E. "John Webster, Tussaud Laureate: The Waxworks in 'The Duchess of Malfi.'" *ELH* 79, no. 4 (2012): 851–877.

Paine, Thomas. *The Rights of Man*. London: J. S. Jordan, 1791.

Parker, E. *A Complete KEY to the New Farce, Call'd "Three Hours after Marriage."* London, 1717.

Pasanek, Brad, and Chad Wellmon. "The Enlightenment Index." *The Eighteenth Century* 56, no. 3 (2015): 359–382.

Paulson, Ronald. *Hogarth*. Vol. 3, *Art and Politics, 1750–1764*. New Brunswick, NJ: Rutgers University Press, 1993.

Payne, Alina Alexandra. Introduction to *Vision and Its Instruments: Art, Science, and Technology in Early Modern Europe*, edited by Alina Alexandra Payne, 1–9. University Park: Pennsylvania State University Press, 2015.

Peover, Michael. "Stained and Painted Glass." In *Horace Walpole's Strawberry Hill*, edited by Michael Snodin and Cynthia Roman, 64–65. New Haven, CT: Yale University Press, 2009.

Peover, Michael, and Kevin Rogers. "New Light on Strawberry Hill: Walpole's Display of Glass and the Representation of the Ancient Abbey." *Journal of Stained Glass* 34 (2010): 8–52.

Pepys, Samuel. *The Diary of Samuel Pepys: A Selection*. Edited by Robert Latham. London: Penguin, 2003.

Petzold, Andreas. "Stained Glass in the Age of Neoclassicism: The Case of Eglington Margaret Pearson." *British Art Journal* 2, no. 1 (2000): 54–60.

Picciotto, Joanna. *Labors of Innocence in Early Modern England*. Cambridge, MA: Harvard University Press, 2010.

Poovey, Mary. *Genres of the Credit Economy: Mediating Value in Eighteenth- and Nineteenth-Century Britain*. Chicago: University of Chicago Press, 2008.

———. *A History of the Modern Fact: Problems of Knowledge in the Sciences of Wealth and Society*. Chicago: University of Chicago Press, 1998.

Pope, Alexander. *The Works of Alexander Pope, Esq., in Verse and Prose: Containing the Principal Notes of Drs. Warburton and Warton*. Edited by William Lisle Bowles, William Warburton, and Joseph Warton. Vols. 3–6. London: J. Johnson, 1806.

Porter, Roy, and Lesley Hall. *The Facts of Life: The Creation of Sexual Knowledge in Britain, 1650–1950*. New Haven, CT: Yale University Press, 1995.

Powell, Thomas. *Humane Industry; or, A History of the Manual Arts, Deducing the Original, Progress, and Improvement of Them*. London, 1661.

Puckett, Kent. *Narrative Theory: A Critical Introduction*. Cambridge: Cambridge University Press, 2016.

Puetz, Anne. "Design Instructions for Artisans in Eighteenth-Century Britain." *Journal of Design History* 12, no. 3 (1999): 217–239.

Pursell, Carroll. "Seeing the Invisible: New Perceptions in the History of Technology." *Icon* 1 (1995): 9–15.

———. "Technologies as Cultural Practice and Production." *Technology and Culture* 51, no. 3 (2010): 715–722.

Rabinbach, Anson. *The Human Motor: Energy, Fatigue, and the Origins of Modernity*. Berkeley: University of California Press, 1992.

Racevskis, Roland. *Time and Ways of Knowing under Louis XIV: Molière, Sévigné, Lafayette*. Lewisburg, PA: Bucknell University Press, 2003.

Rajan, Tilottama. *Dark Interpreter: The Discourse of Romanticism*. Ithaca, NY: Cornell University Press, 1980.

———. "The Work of the Negative: Symbolic, Gothic, and Romantic in Shelley and Hegel." *Studies in Romanticism* 52, no. 1 (2013): 3–32.

Randall, Dale B. J. "Some Observations on the Theme of Chastity in *The Changeling*." *English Literary Renaissance* 14, no. 3 (Autumn 1984): 347–366.

Redfield, Marc. *The Politics of Aesthetics: Nationalism, Gender, Romanticism*. Stanford, CA: Stanford University Press, 2003.

Reeve, Matthew M. "Dickie Bateman and the Gothicization of Old Windsor: Architecture and Sexuality in the Circle of Horace Walpole." *Architectural History* 56 (2013): 97–131.

———. *Gothic Architecture and Sexuality in the Circle of Horace Walpole*. University Park: Pennsylvania State University Press, 2020.

Reeves, Eileen. *Painting the Heavens: Art and Science in the Age of Galileo*. Princeton, NJ: Princeton University Press, 1997.

Reteshka, Joan Rose. "An Edition of Mrs. Hearne's 'The Lover's Week' and 'The Female Deserters.'" PhD diss., Duquesne University, 1998.

Rhodes, Neil, and Jonathan Sawday. "Introduction: Paperworlds: Imagining the Renaissance Computer." In *The Renaissance Computer: Knowledge Technology in the First Age of Print*, edited by Neil Rhodes and Jonathan Sawday, 1–16. London: Routledge, 2000.

Ridley, Mark. *A Short Treatise of Magneticall Bodies and Motions*. London: Nicholas Okes, 1613.

Robbins, John. "Up in the Air: Balloonomania and Scientific Performance." *Eighteenth-Century Studies* 48, no. 4 (2015): 521–538.

Rodgers, Johannah. "The Genealogy of an Image, or, What Does Literature (Not) Have to Do with the History of Computing? Tracing the Sources and Reception of Gulliver's 'Knowledge Engine.'" *Humanities* 6, no. 85 (2017): 1–9.

Rose, Mark. *Authors and Owners: The Invention of Copyright*. Cambridge, MA: Harvard University Press, 1993.

Rosen, William. *The Most Powerful Idea in the World: A Story of Steam, Industry, and Invention*. New York: Random House, 2010.

Rothstein, Eric. *Systems of Order and Inquiry in Later Eighteenth-Century Fiction*. Berkeley: University of California Press, 1975.

Rouquet, Jean Andre. *The Present State of the Arts in England*. London, 1755.

Rousseau, Jean-Jacques. *The Social Contract*. Translated by G.D.H. Cole. New York: E. P. Dutton & Co., 1923.

Russell, Ben. *James Watt: Making the World Anew*. London: Reaktion, 2014.

Ruston, Sharon. *Shelley and Vitality*. London: Palgrave Macmillan, 2009.

Saccaro-Battisti, Giuseppa. "Changing Metaphors of Political Thought." *Journal of the History of Ideas* 44, no. 1 (1983): 31–54.

"Sailing of Bonaparte." *The Morning Post* (London), August 7, 1815.

Salih, Sarah. "Performing Virginity: Sex and Violence in the *Katherine* Group." In *Constructions of Widowhood and Virginity in the Middle Ages*, edited by Cindy L. Carlson and Angela Jane Weisl, 95–112. New York: St. Martin's Press, 1999.

Sampson, Martin W., ed. *Thomas Middleton*. New York: American Book Company, 1915.

Sawday, Jonathan. *Engines of the Imagination: Renaissance Culture and the Rise of the Machine*. London: Routledge, 2007.

Schaffer, Simon. "Enlightened Automata." In *The Sciences in Enlightened Europe*, edited by William Clark, Jan Golinski, and Simon Schaffer, 126–165. Chicago: University of Chicago Press, 1999.

Schmitt, Carl. *Political Theology: Four Chapters on the Concept of Sovereignty*. Translated by George Schwab. Chicago: University of Chicago Press, 2005.

Schonhorn, Manuel. "Defoe's 'Journal of the Plague Year': Topography and Intention." *Review of English Studies*, n.s., 19 (1968): 378–402.

Schwartz, Matthew S. "Huawei Sues U.S. after Congress Bans Government Purchase of Its Equipment." NPR, March 7, 2019. https://www.npr.org/2019/03/07/700989603/huawei-sues-u-s-after-congress-bans-government-purchase-of-its-equipment.

Scott, Walter. Introduction to *The Castle of Otranto* (1811), by Horace Walpole. In *Horace Walpole: The Critical Heritage*, edited by Peter Sabor, 88–98. New York: Routledge, 2007.

Sedgwick, Eve Kosofsky. *The Coherence of Gothic Conventions*. New York: Methuen, 1986.

———. "Paranoid Reading and Reparative Reading, or, You're So Paranoid You Probably Think This Essay Is about You." In *Touching Feeling: Affect, Pedagogy, Performativity*, 123–152. Durham, NC: Duke University Press, 2003.

Serres, Michel. "Turner Translates Carnot." In *Hermes: Literature, Science, Philosophy*, edited and translated by Josué V. Harari and David F. Bell, 55–62. Baltimore: Johns Hopkins University Press, 1982.

Shakespeare, William. *The Tempest*. OpenSource Shakespeare. Accessed July 2019. https://www.opensourceshakespeare.org/views/plays/play_view.php?WorkID=tempest&Act=1&Scene=2&Scope=scene.

Shapin, Steven. *The Scientific Revolution*. Chicago: University of Chicago Press, 1996.

Shapin, Steven, and Simon Schaffer. *Leviathan and the Air-Pump: Hobbes, Boyle, and the Experimental Life*. Princeton, NJ: Princeton University Press, 1985.

Sharp, Jane. *The Midwives Book*. London, 1671.

Shelley, Mary. Preface to *Posthumous Poems of Percy Bysshe Shelley*, iii–viii. London: John and Henry L. Hunt, 1824.

Shelley, Percy Bysshe. *The Letters of Percy Bysshe Shelley*. Vol. 2, *Shelley in Italy*. Edited by Frederick L. Jones. Oxford: Oxford University Press, 1964.

———. *Posthumous Poems of Percy Bysshe Shelley*. London: John and Henry L. Hunt, 1824.

———. *Shelley's Poetry and Prose*. Edited by Donald H. Reiman and Neil Fraistat. New York: Norton, 2002.

Shenstone, William. *The Poetical Works of William Shenstone*. Vol. 1. London, 1780.

Sherburn, George. "The Fortunes and Misfortunes of *Three Hours after Marriage*." *Modern Philology* 24, no. 1 (August 1926): 91–109.

Sherman, Stuart. *Telling Time: Clocks, Diaries, and English Diurnal Form, 1660–1785*. Chicago: University of Chicago Press, 1996.

Shklovsky, Viktor. *Theory of Prose*. Translated by Benjamin Sher. Champaign, IL: Dalkey Archive Press, 1990.

Sielke, Sabine. "Science Studies and Literature." *Anglia* 133, no. 1 (2015): 9–21.

Silver, Sean R. "Visiting Strawberry Hill: Horace Walpole's Gothic Historiography." *Eighteenth Century Fiction* 21, no. 4 (2009): 535–564.

Silverstein, Marc. "'Body-Presence': Cixous's Phenomenology of Theater." *Theatre Journal* 43, no. 4 (December 1991): 507–516.

Simondon, Gilbert. *On the Mode of Existence of Technical Objects*. Translated by Cécile Malaspina and John Rogove. Minneapolis: Univocal, 2017.

Simpson, A.D.C. "Reeve, Richard (d. 1666), Maker of Optical Instruments." In *Oxford Dictionary of National Biography: In Association with the British Academy: From the Earliest Times to the Year 2000*, edited by H. C. G. Matthew and Brian Harrison. Oxford: Oxford University Press, 2004, https://doi.org/10.1093/ref:odnb/48372.

Sinibaldi, Giovanni Benedetto. *Rare Verities: The Cabinet of Venus Unlocked and Her Secrets Laid Open*. London, 1658.

Slack, Paul. *The Invention of Improvement: Information and Material Progress in Seventeenth-Century England*. Oxford: Oxford University Press, 2015.

Sloman, Judith. "The Time Scheme of Defoe's *Roxana*." *English Studies in Canada* 5, no. 4 (1979): 406–419.

Smith, Adam T. *The Political Machine: Assembling Sovereignty in the Bronze Age Caucasus*. Princeton, NJ: Princeton University Press, 2015.

Smith, Bruce D., and Melinda A. Zeder. "The Onset of the Anthropocene." *Anthropocene* 4 (2013): 8–13.

Smith, Courtney Weiss. *Empiricist Devotions: Science, Religion, and Poetry in Early Eighteenth-Century England*. Charlottesville: University of Virginia Press, 2016.

Smith, Pamela H. "Art, Science, and Visual Culture in Early Modern Europe." *Isis* 97 (March 2006): 83–100.

———. *The Body of the Artisan: Art and Experience in the Scientific Revolution*. Chicago: University of Chicago Press, 2004.

Sobel, Dava. *Longitude: The True Story of a Lone Genius Who Solved the Greatest Scientific Problem of His Time*. New York: Walker and Company, 1995.

Solomon, Alex. "The Novel and the Bowling Green: Toby Shandy's Diagrammatic Realism." *Philological Quarterly* 95, no. 2 (2016): 269–291.

Spadafora, David. *The Idea of Progress in Eighteenth-Century Britain*. New Haven, CT: Yale University Press, 1990.

Spiller, Elizabeth. *Science, Reading, and Renaissance Literature: The Art of Making Knowledge, 1580–1670*. Cambridge: Cambridge University Press, 2004.

Sprat, Thomas. "History of the Royal Society." In *English Science, Bacon to Newton*, edited by Brian Vickers, 160–182. Cambridge: Cambridge University Press, 1987.

Stalnaker, Joanna. *Unfinished Enlightenment: Description in the Age of the Encyclopedia*. Ithaca, NY: Cornell University Press, 2010.

Stanyon, Miranda. "Serpentine Sighs: De Quincey's *Suspiria de Profundis* and the Serpentine Line." *Studies in Romanticism* 53, no. 1 (2014): 31–58.

States, Bert O. *Great Reckonings in Little Rooms: On the Phenomenology of Theater*. Berkeley: University of California Press, 1985.

Stauffer, Andrew. *Anger, Revolution, and Romanticism*. Cambridge: Cambridge University Press, 2005.

Sterne, Laurence. *The Life and Opinions of Tristram Shandy, a Gentleman*. London: Penguin, 2003.

St. James's Chronicle or the British Evening Post. September 18, 1794.

Stolberg, Michael. "Examining the Body, *c*. 1500–1750." In *The Routledge History of Sex and the Body*, edited by Sarah Toulalan and Kate Fisher, 91–105. London: Routledge, 2013.

Streeck, Wolfgang. *How Will Capitalism End? Essays on a Failing System*. London: Verso, 2016.

Stuart, Robert. *A Descriptive History of the Steam Engine*. London: Knight and Lacey, 1824.

Sudan, Rajani. *The Alchemy of Empire: Abject Materials and the Technologies of Colonialism*. New York: Fordham University Press, 2016.

Sullivan, Richard J. "Patent Counts and Textile Invention: A Comment on Griffiths, Hunt, and O'Brien." *Journal of Economic History* 55, no. 3 (1995): 666–670.

Summers, David. "*Maniera* and Movement: The *Figura Serpentinata*." *Art Quarterly* 35 (1972): 269–273.

Swartz, Kelly. "The Maxims of Swift's Psychological Fiction." *Eighteenth-Century Fiction* 30, no. 1 (2017): 1–23.

Swift, Jonathan. "A Discourse Concerning the Mechanical Operation of the Spirit." In *The Essential Writings of Jonathan Swift*, edited by Claude Rawson and Ian Higgins, 112–126. New York: Norton, 2010.

———. *Gulliver's Travels*. Edited by Peter Dixon and John Chalker. Baltimore: Penguin, 1967.

———. *Gulliver's Travels*. Edited by Albert J. Rivero. New York: Norton, 2002.

———. *The Prose Works of Jonathan Swift*. Vol. 11. Edited by Herbert Davis. Oxford: Basil Blackwell, 1941.

———. *A Tale of a Tub, to Which Is Added The Battle of the Books and the Mechanical Operation of the Spirit*. 2nd ed. Edited by A. C. Guthkelch and D. Nichol Smith. Oxford, 1968.

———. *The Works of the Rev. Jonathan Swift, DD . . . with Notes, Historical and Critical*. Edited by Thomas Sheridan and John Nichols. Vol. 3. London: J. Johnson, 1808.

Taylor, Jesse Oak. "The Novel after Nature, Nature after the Novel: Richard Jefferies's Anthropocene Romance." *Studies in the Novel* 50, no. 1 (Spring 2018): 108–133.

Teltscher, Kate. *India Inscribed: European and British Writing on India, 1600–1800*. Delhi: Oxford University Press, 1995.

Thomasen, Laura Søvsø, and Henrik Kragh Sørensen. "The Irony of Romantic Mathematics: Bridging the Historiographies of Literature and Mathematics." *Configurations* 24, no. 1 (2016): 53–70.

Thompson, E. P. "Time, Work-Discipline, and Industrial Capitalism." *Past & Present* 38 (1967): 56–97.

Thorne, Christian. "Providence in the Early Novel, or Accident If You Please." *MLQ* 64, no. 3 (2003): 323–347.

Thurschwell, Pamela. *Literature, Technology, and Magical Thinking, 1880–1920*. Cambridge: Cambridge University Press, 2001.

Tierney-Hynes, Rebecca. "Fictional Mechanics: Haywood, Reading, and the Passions." *The Eighteenth Century* 51, nos. 1–2 (Spring–Summer 2010): 153–172.

"Tired of Inequality? One Economist Says It'll Only Get Worse: Interview with Tyler Cowen." NPR, September 12, 2013. http://www.npr.org/templates/transcript/transcript.php?storyId=221425582.

Tullo, Ellen. "Duke Ferdinand: Patient or Possessed? The Reflection of Contemporary Medical Discourse in John Webster's *The Duchess of Malfi*." *Medical Humanities* 36, no. 1 (2010): 19–22.

Turley, Hans. "The Anomalous Fiction of Mary Hearne." *Studies in the Novel* 30, no. 2 (1998): 139–149.

Turner, Henry S. "Lessons from Literature for the Historian of Science (and Vice Versa): Reflections on 'Form.'" *Isis* 101, no. 3 (2010): 578–589.

Turner, Tom. *English Garden Design: History and Styles since 1650*. Suffolk, UK: Antique Collectors' Club, 1986.

Usher, Penelope Meyers. "'I Do Understand Your Inside': The Animal beneath the Skin in Webster's *Duchess of Malfi*.' *Medieval and Renaissance Drama in England* 30 (2017): 105–125.

Venette, Nicolas. *The Mysteries of Conjugal Love Reveal'd*. 2nd ed. London, 1707.

Vermeir, Koen. "The Magic of the Magic Lantern (1660–1700): On Analogical Demonstration and the Visualization of the Invisible." *British Society for the History of Science* 38, no. 2 (2005): 127–159.

Vincent, Clare, and Jan Hendrik Leopold. *European Clocks and Watches in the Metropolitan Museum of Art: Highlights of the Collection*. New York: Metropolitan Museum of Art, 2015.

Vovelle, Michel. *The Fall of the French Monarchy, 1787–1792*. Cambridge: Cambridge University Press, 1984.

Wahrig, Bettina. "Clocks with Hands: Instruments, Hands and Parturients in a Changing Horizon of Time." *History and Philosophy of the Life Sciences* 35, no. 1 (2013): 61–67.

Walpole, Horace. *The Castle of Otranto: A Gothic Story*. Edited by Nick Groom. Oxford: Oxford University Press, 2014.

———. *A Description of the Villa of Horace Walpole, Youngest Son of Sir Robert Walpole Earl of Orford, at Strawberry-Hill, Near Twickenham. With an Inventory of the Furniture, Pictures, Curiosities, &c.* Twickenham, UK: Strawberry Hill Press, 1784.

———. *Horace Walpole's Correspondence*. Edited by W. S. Lewis. 48 vols. New Haven, CT: Yale University Press, 1937–1983.

———. *The Mysterious Mother: A Tragedy*. Twickenham, UK: Strawberry Hill Press, 1758.

Wang, Orrin N. C. *Fantastic Modernity: Dialectical Readings in Romanticism and Theory*. Baltimore: Johns Hopkins University Press, 1996.

Wasserman, Earl. *Shelley: A Critical Reading*. Baltimore: Johns Hopkins University Press, 1977.

Watt, Ian. *The Rise of the Novel: Studies in Defoe, Richardson, and Fielding*. 2nd ed. Berkeley: University of California Press, 2001.

Webster, John. *The Duchess of Malfi*. Edited by Michael Neill. New York: Norton, 2015.

Wecker, Johann Jacob. *Eighteen Books of the Secrets of Art & Nature*. London, 1660.

Weheliye, Alexander G. *Habeas Viscus*. Durham, NC: Duke University Press, 2014.

Weiss, Eric A. "Anecdotes: Jonathan Swift's Computing Invention." *Annals of the History of Computing* 7, no. 2 (1985): 164–165.

Westin, Shannon. "The Obstetric Forceps: Shrouded in Mystery." *Obstetrical and Gynecological Survey* 61, no. 6 (June 2006): 357–358.

Wharton, Joanna. "Maria Edgeworth and the Telegraph." *European Romantic Review* 31, no. 6 (2020): 747–765.

Whitehall Evening Post. September 27, 1794.

Whiston, William. *A New Theory of the Earth, from Its Original, to the Consummation of all Things*. London: Motte, 1737.

Wickman, Matthew. *Literature after Euclid: The Geometric Imagination in the Long Scottish Enlightenment*. Philadelphia: University of Pennsylvania Press, 2016.

Williams, George Walton. Introduction to *The Changeling*, by Thomas Middleton and William Rowley, ix–xxiv. Edited by George Walton Williams. Lincoln: University of Nebraska Press, 1966.

Williams, Raymond. *Marxism and Literature*. Oxford: Oxford University Press, 1977.

Williams, Rosalind. "The Lantern-Bearers of the History of Technology." *History and Technology* 29, no. 3 (2013): 262–277.

Wilson, Adrian. *The Making of Man-Midwifery: Childbirth in England, 1660–1770*. Cambridge, MA: Harvard University Press, 1995.

Wilson, Geoffrey. *The Old Telegraphs*. London: Philimore, 1976.

Wilson, Ross. *Shelley and the Apprehension of Life*. Cambridge: Cambridge University Press, 2013.

Winton, Calhoun. *John Gay and the London Theatre*. Lexington: University Press of Kentucky, 1993.

Witmore, Michael. "Arrow, Acrobat and Phoenix: On Sense and Motion in English Civic Pageantry." In *The Automaton in English Renaissance Literature*, edited by Wendy Beth Hyman, 109–127. Burlington, VT: Ashgate, 2011.

Wolfe, Jessica. *Humanism, Machinery, and Renaissance Literature*. Cambridge: Cambridge University Press, 2004.

Wolfson, Susan. *Formal Charges: The Shaping of Poetry in British Romanticism*. Stanford, CA: Stanford University Press, 1997.

Wordsworth, William. *Lyrical Ballads, and Other Poems, 1797–1800*. Ithaca, NY: Cornell University Press, 1993.

Worshipful Company of Clockmakers. *The Clockmakers' Farther Reasons against Mr. Hutchinson's Bill and Printed Reasons*. London, 1712.

———. *Reasons Humbly Offered . . . against the Bill, Entituled, a Bill for Securing to Mr. John Hutchinson the Property of a Movement Invented by Him for the More Exact Measuring of Time*. London, 1712.

Yahav, Amit S. *Feeling Time: Duration, the Novel, and Eighteenth-Century Sensibility*. Philadelphia: University of Pennsylvania Press, 2018.

———. "Time, Duration, and Defoe's Novels." *Partial Answers: Journal of Literature and the History of Ideas* 6, no. 1 (2008): 33–56.

Yang, Chi-ming. "Asia Out of Place: The Aesthetics of Incorruptibility in Behn's *Oroonoko*." *Eighteenth-Century Studies* 42, no. 2 (2009): 235–253.

Ylivuori, Soile. "Rethinking Female Chastity and Gentlewomen's Honour in Eighteenth-Century England." *The Historical Journal* 59, no. 1 (March 2016): 71–97.

Young, Helen. "Race, Medievalism and the Eighteenth-Century Gothic Turn." *Postmedieval: A Journal of Medieval Cultural Studies* 11, no. 4 (2020): 468–475.

Zitin, Abigail. *Practical Form: Abstraction, Technique, and Beauty in Eighteenth-Century Aesthetics*. New Haven, CT: Yale University Press, 2020.

Zuroski, Eugenia. "'That You May Show Us What We Have Seen': Bentley's Drawings and the Archive of Queer Feeling." Paper presented at "Text Artefact Identity: Horace Walpole and the Queer Eighteenth Century," Twickenham, UK, February 2019.

Notes on Contributors

JOSEPH DRURY is an associate professor of English at Villanova University in Villanova, Pennsylvania. He is the author of *Novel Machines: Technology and Narrative Form in Enlightenment Britain*, as well as several articles on the intersections of eighteenth-century literature, science, and technology.

LAURA FRANCIS is a PhD candidate in the Department of Literatures in English at Cornell University in Ithaca, New York. She specializes in sixteenth- and seventeenth-century writing, with a particular focus on translation theory, baroque aesthetics, and the influence of Spanish culture in England. While her dissertation explores how the process of translation shaped the expansion of early modern empires into the worlds around them, she maintains a keen interest in specifically scientific exchanges through her other presentations and work in the digital humanities. Her research has been supported by the Government of Ireland, the Mario Einaudi Center for International Studies at Cornell University, and the Cornell Society for the Humanities.

KRISTIN M. GIRTEN is an associate professor of English and assistant vice chancellor for the arts and humanities at the University of Nebraska at Omaha. Her research focuses on intersections between literature, philosophy, and science in the British Enlightenment, giving special emphasis to how women and other marginalized groups contribute to and feel the effects of such intersections. Her first book, currently in progress, establishes an alternative history of knowledge making in the long Enlightenment era by tracing feminist practices of empirical witnessing across the period. Her essays have appeared in *Eighteenth-Century Studies* and *The Eighteenth Century: Theory and Interpretation*.

AARON R. HANLON is an associate professor of English and chair of the Science, Technology, and Society program at Colby College in Waterville, Maine. He is the author of *A World of Disorderly Notions: Quixote and the Logic of Exceptionalism*.

ERIK L. JOHNSON is a lecturer in the humanities department at San José State University in California, where he teaches in the school's Humanities Honors program. His publications include "Life beyond Life: Reading Milton's *Areopagitica* through Enlightenment Vitalism" in *Eighteenth-Century Studies* and pieces on teaching William Congreve and William Wycherley in the collection *How to Teach a Play: Essential Exercises for Popular Plays*, edited by Miriam Chirico and Kelly Younger.

JAMISON KANTOR is an assistant professor of eighteenth- and nineteenth-century literature at the Ohio State University. His first book, *Honor, Romanticism, and the Hidden Value of Modernity*, pits the economized and oftentimes reactionary concept of "liberty" against honor, a surprisingly radical social idea found in many cultural forms, from landscape poetry to the media of the Black diaspora. His second book, currently in progress, expands this critique of rationalist value to structures of automation—and "automatic materiality"—found in literature and art from 1750 to 1850. Kantor's essays have also appeared in *The Eighteenth Century: Theory and Interpretation*, *PMLA*, and *Jump Cut*.

KEVIN MACDONNELL is an assistant professor of humanities at Massachusetts Maritime Academy in Buzzards Bay, Massachusetts. His book project, "Genres of Innovation: Empire, Environment, and Economy in the Eighteenth-Century British Atlantic," examines how eighteenth-century literature shaped the reception and application of the innovations that would marshal the onset of British hegemony in the Atlantic world. His most recent publications can be found in *Philological Quarterly* and *Green Theory & Praxis*.

ZACHARY M. MANN is a doctoral candidate in English literature at the University of Southern California in Los Angeles. He has published on new media, television, and film history, most recently on the career of Myrna Loy. His dissertation traces the co-evolutions of machine programming and conceptions of authorship from the eighteenth century to today.

THOMAS A. OLDHAM is an assistant professor of theatre studies at Texas A&M University–Corpus Christi. Recent publications include "Philip Ridley: Still In-Yer-Face" in the edited collection *After In-Yer-Face: Remnants of a Theatrical Revolution* and "The Affective Appeal of Violence and the Violent Appeal of Affect: *Titus Andronicus*, Lucy Bailey, and Shakespeare's Globe" in the journal *Shakespeare Bulletin*. He previously taught at Colby College, Indiana University, and Indiana State University, and holds dramaturgy credits from professional and educational theaters in Nebraska, Maine, Indiana, and New York.

DEVEN M. PARKER is a postdoctoral research associate in the School of English and Drama at Queen Mary University of London, where she works on the British Academy project "Romantic Melodrama: Feeling in Search of Form." She specializes in eighteenth- and nineteenth-century literature, theater, and media history. Her most recent publications include "Precarious Correspondence in *The Woman of Colour*" in *Essays in Romanticism* and "Romanticism's Radical Connectivity" in *European Romantic Review*.

Emily M. West is an assistant professor in the Department of English at the University of Hawai'i at Mānoa. Her work focuses on embodiment and material culture in the long eighteenth century. She has recently published on penmanship in "Technologies of Epistolarity and Embodiment in John Cleland's *Memoirs of a Woman of Pleasure*" in *The Eighteenth Century: Theory and Interpretation* and on pedagogical toys in "Animal Things, Human Language, and Children's Education" in *Eighteenth-Century Fiction*. She is currently completing a book project on technology and gendered embodiment.

Index

Note: Page numbers in italics indicate figures.

Kolibri 360 GmbH
Ferdinand-Jühlke-Straße 7
99095 Erfurt, Deutschland
produktsicherheit@kolibri360.de